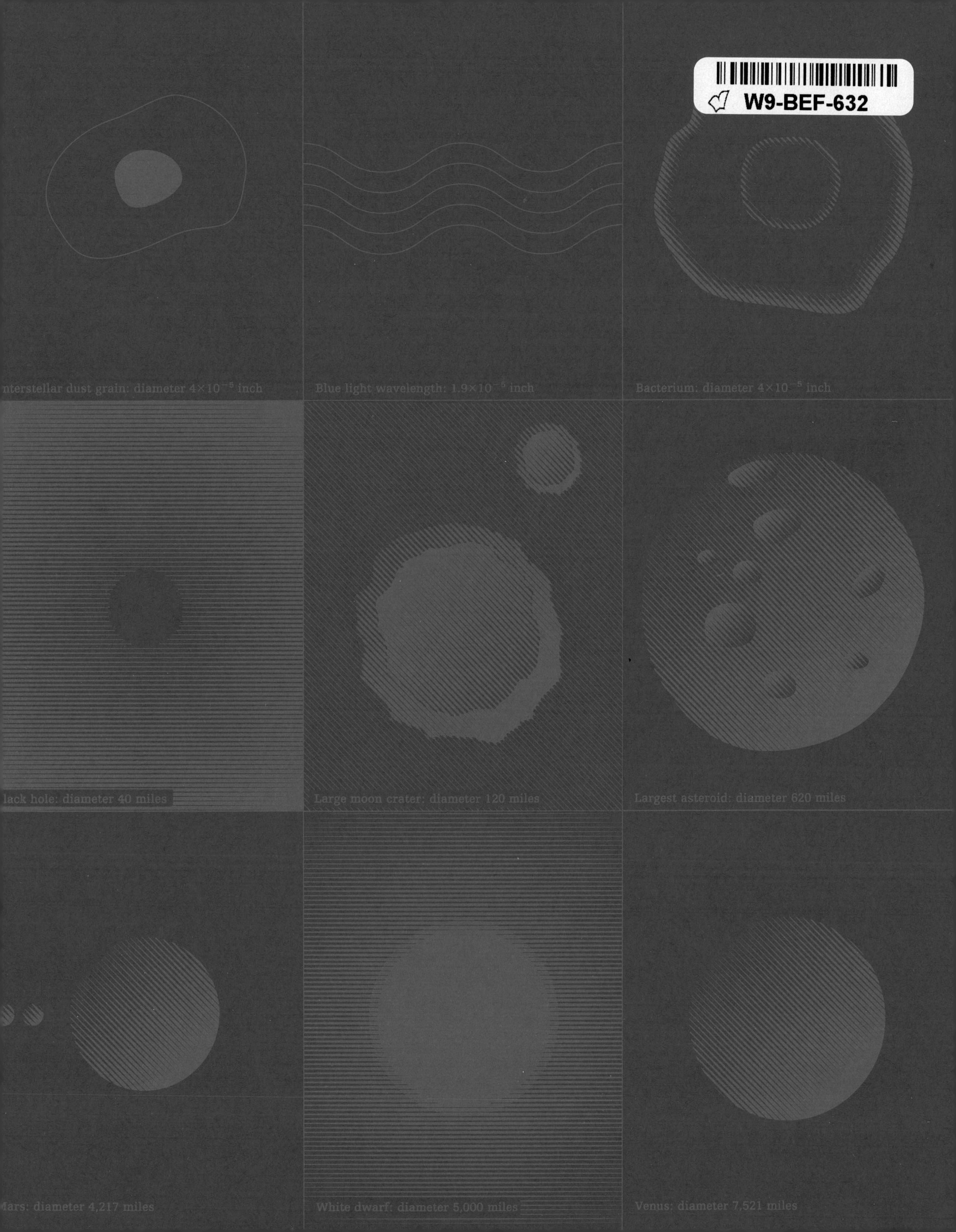
W9-BEF-632
nterstellar dust grain: diameter 4×10⁻⁵ inch
Blue light wavelength: 1.9×10⁻⁵ inch
Bacterium: diameter 4×10⁻⁵ inch
lack hole: diameter 40 miles
Large moon crater: diameter 120 miles
Largest asteroid: diameter 620 miles
Mars: diameter 4,217 miles
White dwarf: diameter 5,000 miles
Venus: diameter 7,521 miles

LIFE SEARCH

A symbolic ship of carbon atoms, the basic building blocks of all known life, sails toward a cluster of young stars cocooned in dust and gases.

TIME
LIFE ®

Other Publications:
AMERICAN COUNTRY
THE THIRD REICH
THE TIME-LIFE GARDENER'S GUIDE
MYSTERIES OF THE UNKNOWN
TIME FRAME
FIX IT YOURSELF
FITNESS, HEALTH & NUTRITION
SUCCESSFUL PARENTING
HEALTHY HOME COOKING
UNDERSTANDING COMPUTERS
LIBRARY OF NATIONS
THE ENCHANTED WORLD
THE KODAK LIBRARY OF CREATIVE PHOTOGRAPHY
GREAT MEALS IN MINUTES
THE CIVIL WAR
PLANET EARTH
COLLECTOR'S LIBRARY OF THE CIVIL WAR
THE EPIC OF FLIGHT
THE GOOD COOK
WORLD WAR II
HOME REPAIR AND IMPROVEMENT
THE OLD WEST

This volume is one of a series that
examines the universe in all its aspects,
from its beginnings in the Big Bang to the
promise of space exploration.

VOYAGE THROUGH THE UNIVERSE

LIFE SEARCH

BY THE EDITORS OF TIME-LIFE BOOKS
ALEXANDRIA, VIRGINIA

CONTENTS

1/Reconnaissance

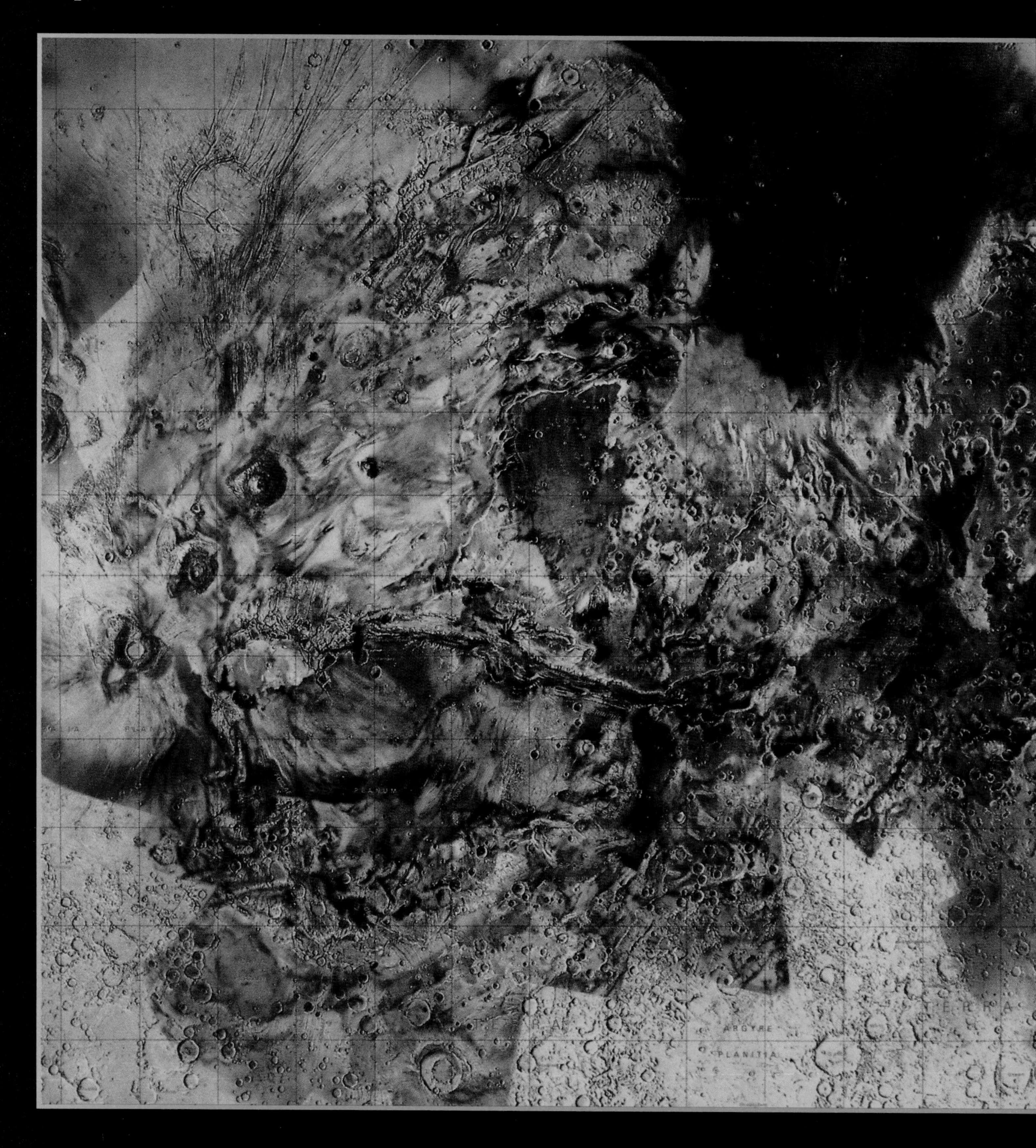

Constructed from Viking images, this U.S. Geological Survey map shows the rough, ruddy face of equatorial Mars—terrain that evokes an earlier epoch when abundant liquid water etched great canyons on the Red Planet.

uiet as the boulders surrounding them, the two scouts from Earth stand on plains of russet sand and rocky debris that stretch to every horizon before shading into a salmon-colored sky. Ultraviolet radiation and dust storms have scrubbed their metal flanks and bleached their once-festive insignias—a tricolored flag, a stylized star, an emblem bearing a Norseman's horned helmet. Their now-feeble nuclear power sources continue to decay; their electronics are inert. But before time and toil caught up with them, the robotic visitors performed with a valor worthy of their shared name: Viking. In 1976 one touched down in the northern hemisphere of Mars, in an area named Chryse Planitia, the other at Utopia Planitia, 4,600 miles and nearly a hemisphere to the west. And at these bridgeheads to a new world, they carried out their mission until November 5, 1982, when the lander at Chryse transmitted the last of the pair's images to Earth: rocks, smooth drifts of rusty sediments, a shallow trench dug by the little scout some years before. Now as silent and blind as the stone circles and cairns erected by the Vikings of another time and place, the twin landers carry a message for all who follow: On this ground began Earth's first direct search for life on other worlds.

The next wave of spacefarers—more robot surrogates at first, but followed very soon by their human masters—will return to Mars to extend the quest begun by Viking, eventually pressing beyond Mars toward other bodies within the Solar System. But even as those preparations get under way, the search is crossing the boundary of the solar neighborhood to probe the star-peppered wilderness of the Milky Way galaxy. Modern searchers listen to the naturally generated hiss and snarl of radio energy raining from the sky, hoping to hear the signals of other civilizations light-years away. With sophisticated telescopes sensitive to all forms of electromagnetic radiation, astronomers rake the heavens for the tiniest hints of nonsolar planetary systems: Where such systems exist, there might be livable worlds.

But what—out in the galaxy or close to home—qualifies as a livable world? Definitions of livability are elusive. The upper atmosphere of gaseous Jupiter, for example, might conceivably support some buoyant life form, and living things might ply the ice-covered waters of the Jovian satellite Europa.

Beneath the orange smog of Titan, Saturn's giant moon, may lie methane seas and continents of crude organic material that could someday give rise to creation. Even in the virtually sunless gloom on the very rim of the Solar System, Triton, the larger of Neptune's two moons, may harbor life in chilled oceans of liquid nitrogen.

Inhospitable as these environments may appear, they cannot be ruled out. Over and over again, incredulous explorers have discovered earthly organisms not only enduring but thriving under seemingly impossible conditions. Organisms make do with miserly rations of sunlight and warmth in the wastelands of Antarctica; bacteria flourish in the radioactive wreckage of damaged nuclear reactors; and the sunless abysses of the sea harbor diverse, prospering colonies of creatures large and small. Clearly, life does what it must to survive, and visionary scientists have speculated on a host of extraterrestrial survival mechanisms: biologies based on silicon instead of carbon; intelligence that exists as pure energy; beings whose bodies are the tenuous hot plasmas of a star, with lives reckoned in billionths of a second.

The belief that other worlds may hold some form of life—that they may be home to another intelligence—is an ancient human dream, one composed about equally of humility and arrogance, belief and heresy, science and myth. This persistent human hope carries a heavy philosophical burden as well, requiring that such imponderables as life and intelligence be described and somehow quantified. Some scientist-philosophers believe that life is everywhere in the universe and may be accessible over the daunting distances separating stars. Others surmise that this blue planet near the Sun is life's sole point of origin and that the universe is as lonely as it looks, a vast desert only humans will explore. Still, the dream persists.

CANALI

Perhaps inevitably, the longing to prove that mankind is not alone in the universe came to focus most intensely upon a world that seemed remarkably Earth-like: Mars. As the advent of telescopes allowed astronomers to study the planet and its habits more closely, various similarities appeared. Although its year is nearly twice as long as Earth's, Mars's twenty-four-and-a-half-hour day seems to mimic its terrestrial counterpart. Dark regions on the globe wax and wane seasonally, evoking oceans, continents, and vegetation; the Martian polar ice caps advance and retreat before the Sun.

A century or so ago, the conviction that Mars sustains life was fueled by such tantalizing views through the telescope. At its most distant from Earth—some 248 million miles—Mars is merely a featureless orange disk not much larger than a star. But every 780 days, when Mars and Earth line up together on the same side of the Sun (a configuration known as opposition because Mars and the Sun are opposite one another in Earth's sky), the Red Planet can acquire a distinctive personality. This is especially true on the rare occasions, every fifteen to nineteen years, when Mars reaches opposition at the same time that it reaches perihelion, the point in its elliptical orbit where it is closest to the Sun. During these perihelic oppositions, Mars swings within 35 million miles of Earth, and astronomers rush to their instruments for an observational feast.

The perihelic opposition of 1877 was just such a banquet for Giovanni Schiaparelli, an accomplished astronomer and the director of Milan's Brera Observatory. Working at the eyepiece of his telescope, Schiaparelli would focus on one feature of the planet at a time, looking away to make a meticulous drawing, then returning to the eyepiece. In this fashion he compiled a sketch of what he saw. Gradually, an astonishing pattern emerged: Mars was covered by a planet-spanning network of geometrically ordered lines. He was not the first to see lines on the Martian surface or to see the hand of intelligence in their neat geometry. But Schiaparelli's report staggered the imagination. He called the lines *canali,* Italian for "channels," or canals. In a time when the 107-mile-long Suez Canal, completed less than a decade earlier, was still cause for wonder, the public inevitably took *canali* to mean canals and began to imagine beings capable of digging them.

The archetype of a turn-of-the-century gentleman astronomer, Percival Lowell adjusts the twenty-four-inch refracting telescope, for years the primary vehicle of his search for the canals of Mars and the beings who built them. Installed on Mars Hill west of Flagstaff, Arizona, in 1896, the instrument was then among the world's most powerful telescopes.

Such imaginings received immediate help from Paris, where Camille Flammarion, one of history's great scientific speculators, gave quick support to Schiaparelli's discovery. Some fifteen years earlier, Flammarion, then a twenty-year-old French writer-astronomer, had published *La Pluralité des Mondes Habités*—the plurality of inhabited worlds. Written in a flamboyant style that would become his trademark, the book made Flammarion a kind of sage on the subject of extraterrestrial life in general and life on Mars in particular. By the time of the 1877 Mars opposition, he had become a noted popularizer of science, and his ready acceptance of Schiaparelli's report served to reinforce the general belief in intelligent life on the fourth planet from the Sun. The influence of Schiaparelli and Flammarion was not limited to European soil. Across the Atlantic, an American named Percival Lowell was also feeling the tug of Mars and its canal builders.

For most of his adult life, Lowell would have been recognized chiefly as a son of the prosperous Lowell family of Boston and as an oriental scholar of

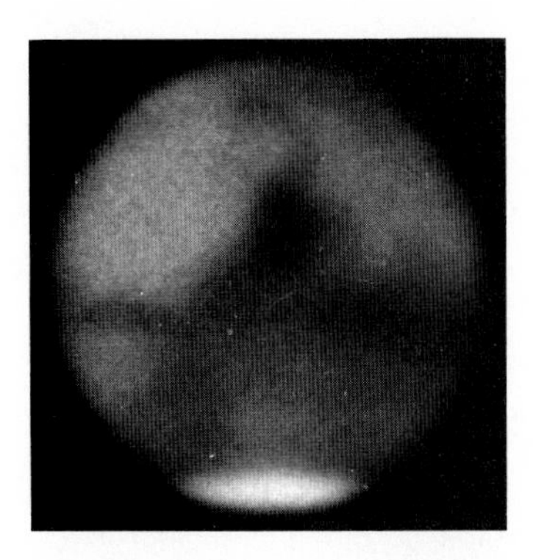
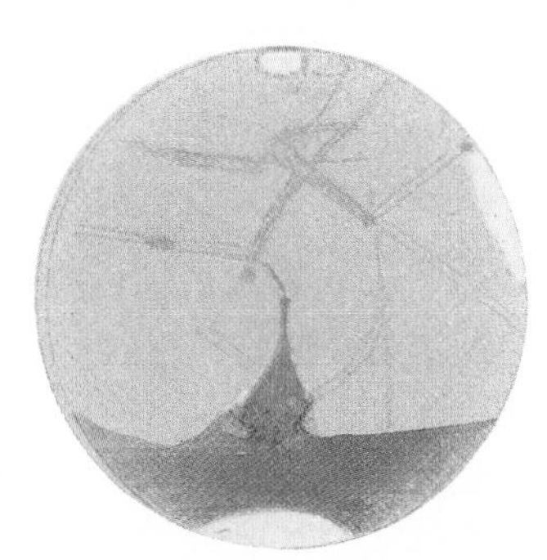
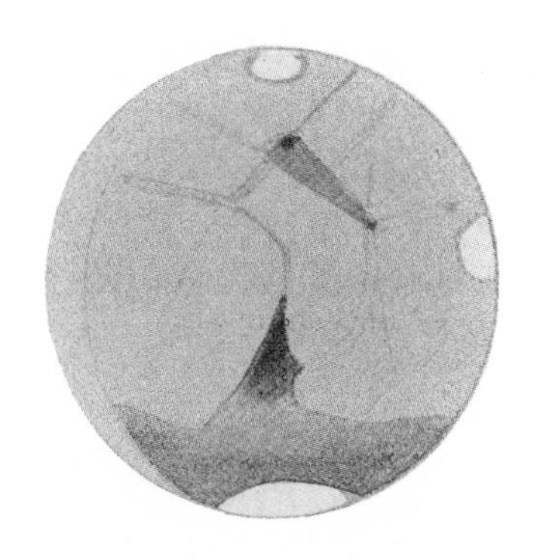
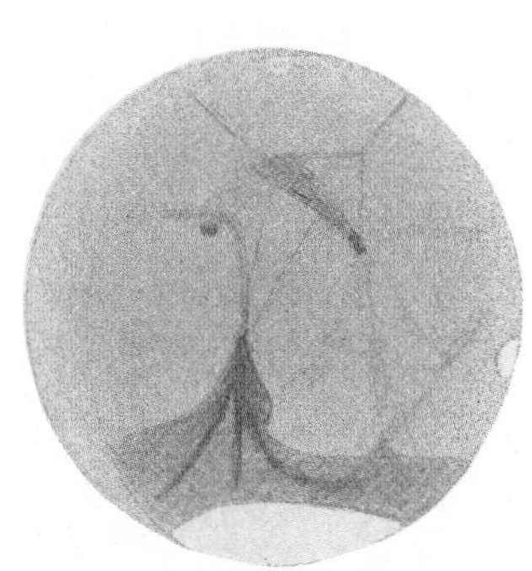

To Lowell, and to many colleagues, the rough Martian landscape resolved itself into lines that were believed to be a network of canals. This landscape Lowell rendered in delicately shaded sketches like the three shown above right, colored to show vegetation *(blue-green)*, desert *(tan)*, and fallow land *(brown)*. In 1907 Lowell turned to photography for final proof that the canals existed. But even the relatively sharp images produced by the new technology *(above)* could not bring the canals into undisputed focus.

some standing. But in 1893 he returned from his final journey to the East and wrote his last book on the subject. Then, at thirty-nine, he unaccountably turned to astronomy and to Mars, publishing an article extolling the canals as "the work of some sort of intelligent beings." By the spring of 1894—just in time for the next Martian perihelic opposition—he had established an observatory near Flagstaff, Arizona, on a pine-sprinkled mesa where he would pursue the elusive inhabitants of the Red Planet off and on for the rest of his life *(page 17)*. After less than a year of formal observation, he announced his views to the world in a book called *Mars*.

The Martians envisioned by Lowell were a serious race bound in peace and common effort by the relentless desertification of their planet. "Irrigation," he wrote, "must be the chief material concern of their lives." Moreover, the complexity of their global network of canals and oases evoked intellects superior to humans'. Lowell's Mars had to be dry enough to require extensive irrigation but wet enough for water to gush through the vast system of canals as the polar caps took turns melting and refreezing. The Martian atmosphere, as Lowell conceived it, contributed to this climatic state. He decided that because Mars is only about one-tenth as massive as Earth, its weaker gravitational field allowed much of the atmosphere to escape into space, leaving a residue about one-seventh the density of Earth's atmosphere—thick enough for water vapor to float suspended in it, but not so thick that the vapor could condense into rain.

THE WRECKING CREW

A counterattack began immediately after the book's release. Although the publication of *Mars* generated wide interest, it was scorned by many in the scientific community, who believed Lowell's claims were entirely unsupported. For example, William Wallace Campbell, a thirty-three-year-old astronomer at California's Lick Observatory, gave the book a crushing review. "Mr. Lowell," said Campbell, "went direct from the lecture hall to his observatory, and how well his observations established his pre-observational views is told in his book." Campbell's opinion of Lowell himself was even lower. "He has taken the popular side of the most popular scientific question about," Campbell wrote. "The world at large is anxious for the discovery of intelligent life on Mars, and every advocate gets an instant and huge audience."

Hard as Campbell was on the man, his own work proved even harder on the Mars that Lowell had concocted. During the opposition of 1894—even before *Mars* was published—Campbell had set out to reconfirm earlier observations of water in the Martian atmosphere. Because similar measurements had been made by at least four other observers, Campbell actually expected the task to be "a simple and easy matter." He was wrong. After repeating his observations ten times, he was finally confident of his results: His data, he reported, showed "no evidence whatever" of water vapor in the Martian atmosphere. Without water, there could be no irrigation; without irrigation, there was no reason for canals or for Martian superengineers to build them.

The necessity of irrigation aside, some scientists questioned the objective reality of what Lowell and many other astronomers had seen. For example, shortly after the turn of the century, British astronomer Edward Walter Maunder devised a novel experiment to prove how deceiving observation can be. In thirteen tests, groups of about twenty teenage boys at the Greenwich Royal Hospital School were shown drawings in which Mars was simply a delicately shaded disk. The majority of the boys, drawing what they saw, produced a Mars covered with geometrical lines *(opposite).*

The subtle arguments of what came to be called the illusionist position were largely lost on a public still convinced that Lowell's Mars was real. But in 1909, the year of the next perihelic Martian opposition, one influential believer in the Martian canals changed his mind.

The Turkish-born astronomer Eugene Antoniadi had succeeded Edward Maunder as director of the Mars section of the British Astronomical Association in 1896. Soon he was considered one of the world's experts on the Red Planet. Henri Deslandres, director of the Paris Observatory, invited him to spend the autumn of 1909 at the observatory's station in Meudon, housed in a seventeenth-century château above the Seine; the facility possessed a superb 32.7-inch refractor telescope. This was the tenth time Antoniadi had observed Mars in opposition, but the combination of perihelic opposition and a powerful viewing instrument told a new story. Where once he had seen lines, he now saw a chaotic smear of dark and light areas. "It was at once obvious," he wrote of the experience, "that the geometrical network of single and double canals discovered by Schiaparelli was a gross illusion."

Despite the increasing body of contrary evidence from scientists like Maunder, Campbell, and Antoniadi, Lowell's belief never wavered. Until his death in 1916, he continued to describe the gallant inhabitants of Mars, courageously engaged in a war they could not win. "The process that brought it to its present pass must go on to the bitter end," he wrote in 1908, "until the last spark of Martian life goes out. When the last ember is thus extinguished, the planet will roll a dead world in space, its evolutionary career forever ended."

Believers in a Martian civilization did not disappear with Lowell's passing. In the late 1950s, for example, the discovery of slight but inexplicable per-

Dismissed as "the small boy theory" by a defensive Percival Lowell, an experiment designed in 1903 by British astronomer Edward Maunder did much to establish that the Martian canals were an optical illusion caused by the eye's tendency to combine fine detail. In his tests, Maunder showed groups of teenage Greenwich boys a key map in which Mars appeared as a shaded blur of detail *(opposite, top).* The boys, studying the sketch from various distances, almost invariably produced a planetary disk festooned with networks of canal-like lines *(opposite, below).*

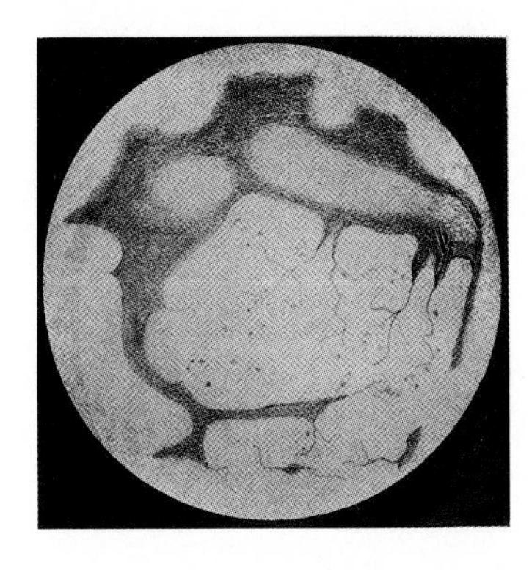

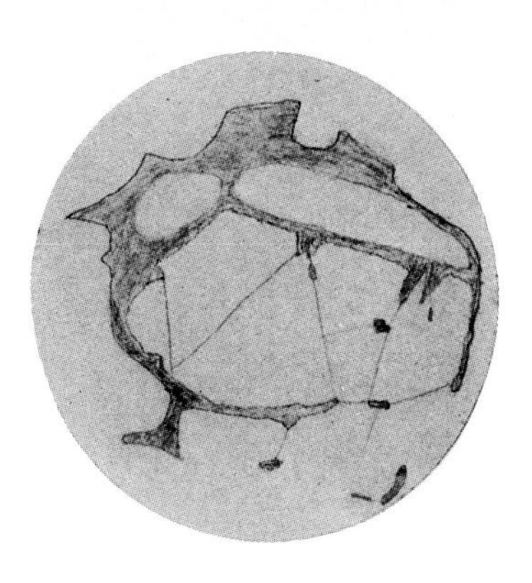

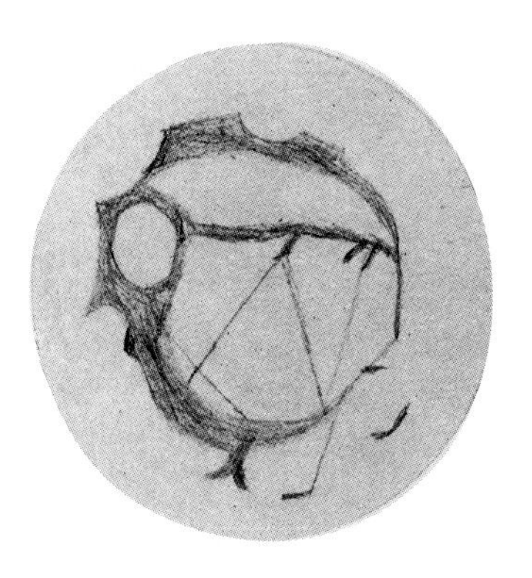

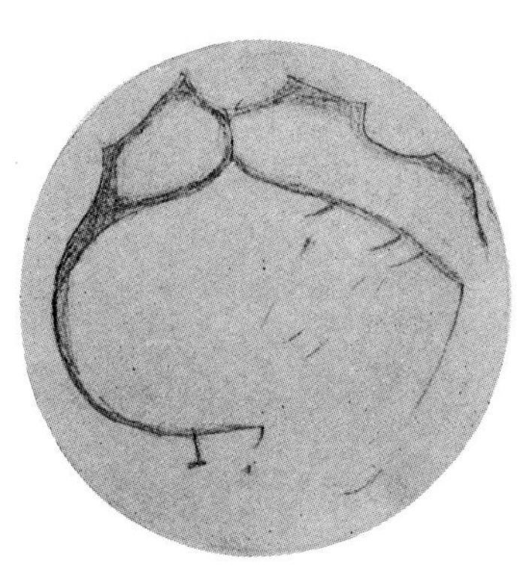

turbations in the orbits of the two Martian moons, Phobos and Deimos, prompted some provocative solutions, including one by Iosef Shmuelovich Shklovskii, director of radio astronomy at the Sternberg Astronomical Institute of the Soviet Academy of Sciences. "Only one possibility remains," he wrote in 1959. "We are led to the possibility that Phobos—and possibly Deimos as well—may be artificial satellites of Mars." They could, he pointed out, be either present-day artifacts of an advanced civilization or the relics of a vanished society.

But as space scientists began the countdown to humankind's first close look at Mars, the days of such flights of fancy were numbered. In 1965 *Mariner 4* sent back twenty-one images of the planet. Although of poor resolution, they showed it to be a desert world splattered with craters, like the Moon. Four years later, the 200 pictures transmitted by *Mariner 6* hinted at a smoother Mars but still could not fully resolve the detail. *Mariner 9,* which went into Martian orbit in 1971, photographed the planet for nearly a year. Its thousands of images, covering virtually the entire globe, revealed the rugged terrain and varied coloration that had given rise to the Lowellian illusion but turned up no sign of canals or the beings who might have built them. One of the more significant findings of *Mariner 9* was that the seasonal changes in the planet's shadings—once perceived as oceans or vast reaches of vegetation—were actually the work of huge Martian dust storms that changed the patterns of reflected sunlight.

Despite these deflating revelations, those who still believed in Martian life took comfort in the scraps of hope *Mariner 9* offered. Unlike the dead Moon, Mars everywhere showed signs of geologic vitality: enormous fissures that dwarfed the Grand Canyon, volcanoes many times the size of Mount Everest, and a face scarred by crustal splitting and what appeared to be the erosive handiwork of wind and water. Although biologists had greatly lowered their expectations for what could live on Mars, surely a planet so vividly marked by its geology would offer something for them as well—perhaps microbial life forms, or at least evidence that such life had once existed in an earlier, wetter epoch.

And science was ready to take up the hunt. Even as *Mariner 4* was making its flyby in 1965, planetary astronomers, geophysicists, and biologists were sketching the objectives and logistics of landing a robot biochemistry laboratory on the Martian surface. By late 1968, when *Mariner 6* was being readied for launch from Kennedy Space Center, the National Aeronautics and Space Administration had begun to design Viking, the first real life-search vehicle.

MICROBE HUNTERS

The task of shaping Viking's scientific mission went to Gerald A. Soffen, a biologist from the Jet Propulsion Laboratory in Pasadena, California. In 1968, when the forty-two-year-old Soffen moved to NASA and the Viking project, he was taking a calculated risk. In those days, as Stanford Nobel laureate Joshua Lederberg explained much later, one did not lightly opt for a career in exobiology, as the study of extraterrestrial life is known. Lederberg himself had such stature in the scientific world that he could afford to play what he called "a nonreputable game." Soffen lacked the safety net of fame. Nevertheless, for eight years at JPL he worked on developing life detectors and automated biology laboratories for the Mars missions that ultimately crystallized into Viking.

The instrument visualized by Soffen's science team (which included Lederberg) became more complex as it took form *(pages 20-24).* Although some scientists had urged NASA not to be what they called geochauvinistic, Viking's biology package was shaped by the assumptions that Martian life would resemble terrestrial life and could be recognized with the same techniques. Biologist Frederick S. Brown, who was a Viking project scientist at TRW, Inc., the firm that designed the biology instrument, defended the decision in terms of scientific experience. "After decades of investigation," he said before the mission, "we have no reason to think life—life anywhere—would evolve with anything but carbon biochemistry. It would beat out other less complex and adaptable biochemical systems on an evolutionary level."

Viking's final biology package incorporated three separate life-detection schemes, two of them employing radioactive carbon 14 as a tracer for biological responses. One of the tracer experiments, designed by Norman Horowitz of the California Institute of Technology, would look for evidence that Martian organisms took up, or fixed, the carbon in carbon dioxide, as terrestrial plants do in the course of growing. Termed the pyrolytic-release experiment, it was the only one aboard Viking that attempted to simulate actual conditions on Mars. The other tracer scheme was the labeled-release experiment. Proposed by Gilbert Levin of a Maryland firm called Biospherics Research, Inc., it would seek signs of metabolism—the transformation of nutrients into energy—by measuring how much radioactively labeled carbon dioxide an incubating sample released into a sealed test chamber. A third experiment, designed by Vance Oyama, a life-detection specialist at NASA's Ames Research Center, would sense respiration by analyzing the gases released by an incubating sample as it received moisture and nutrients over periods of days and months.

Because all known life, past and present, creates an abundance of organic, or carbon-based, compounds *(pages 29-39),* the lander would also carry an automated, miniaturized combination of two of the most elaborate analytical tools in earthbound laboratories: a gas chromatograph and a mass spectrometer *(page 24).* Developing the instrument for Viking proved more difficult than anticipated. The two-headed device became the target of efforts to simplify it nearly out of existence or replace it with another design that did not use a gas chromatograph. But Soffen, realizing that both instruments were needed to determine whether the soil of Mars contained organic substances, successfully fought to keep the combined gas chromatograph/mass spectrometer, or GCMS, on Viking.

On August 20, 1975, the first Viking probe rode a Titan III Centaur rocket

The View from Mars Hill

Convinced of the existence of a noble, canal-building race on Mars, Percival Lowell set out in 1893 to build an observatory that could provide scientific proof. If his motives were dubious, his technical standards for the scouting of other worlds were ahead of his time. Lowell was among the first to insist on a high, dry location that offered "excellence of seeing," noting that "a steady atmosphere is essential to the study of planetary detail."

After testing several candidate sites in the American Southwest, he chose a mountain near Flagstaff, Arizona, for its splendid viewing conditions. Built on what came to be called Mars Hill and equipped with twelve- and eighteen-inch refractors, Lowell Observatory began operations on May 31, 1894, just in time for that year's favorable alignment, or opposition, of Mars and Earth.

Work at the facility was not entirely canals and Martians. Besides pioneering photographic and spectrographic studies of Mars, Lowell and his staff observed other planets and calculated the location of a hypothetical Planet X beyond Neptune, which led to the discovery of Pluto in 1930 by Clyde Tombaugh, who was then a young assistant at the observatory.

When Lowell died in 1916, his will provided the astronomical center with a permanent endowment. However, the Lowell Observatory would long suffer from another legacy: the scientific stigma conferred by the founder's Martian adventures. Even the discovery of Pluto was diminished by the Lowell brand. Tombaugh admitted to feeling "like an outcast in professional circles" and noted that the observatory's senior staff members were effectively ostracized by their colleagues in the scientific field.

The questionable origins of the observatory cast a pall over the entire field of planetary studies and prompted many of its practitioners to shift to more rewarding work in stellar astronomy. But time has repaired the damage. Today, with eight telescopes, Lowell Observatory has evolved into a major astronomical facility, its research divided about equally between Solar System and extragalactic astronomy. And the strong planetary tradition endures, sustained by such successes as the 1988 detection of an unexpectedly dense atmosphere on Pluto, the planet discovered by the house that Martians built.

into the Florida skies. Three weeks later, on September 9, its twin headed for the Red Planet. After centuries of human speculation, the search for Martian life was under way.

MARTIAN SIGNALS

Each Viking ship was in fact a dual spacecraft—an orbiter that would circle Mars and a lander that would actually touch down on the surface. Carefully sterilized to prevent the contamination of Mars by earthly microbes, the landers reposed atop the orbiters in a streamlined pod called an aeroshell, sleeping with metal arms and legs tucked against the body, their electronics dimmed. Once clear of the Earth's atmosphere, the orbiters opened out their solar panels and locked their navigation systems onto Canopus, the second-brightest star in Earth's sky. The next ten months would be a lonely 200-million-mile chase that would take the two Vikings halfway around the Sun to their planetary rendezvous.

Mars was clutched in its deep-frozen winter when the probes began their lengthy journey. By the time the first Viking overtook the planet, frost patches and chilly mists had burned away beneath an advancing summer Sun. In the early part of July, with the northern half of Mars in summer, the first Viking probe entered a highly eccentric orbit around the planet and prepared to launch its lander.

The landing sites had been selected from low-resolution *Mariner 9* images. Now, as Viking orbited Mars, its more advanced cameras showed that what had looked like inviting fields were in fact a jumble of canyons and craters. Mission managers at JPL put the lander launch on hold and redirected the probe into a month's reconnaissance from orbit before agreeing on a less-hazardous landing point: a broad swatch of level ground that Schiaparelli had named Chryse Planitia, the Golden Plain, ninety-nine years before.

On July 20, 1976, *Viking 1* rotated to aim its lander toward Mars. As the pod began its three-and-a-half-hour fall to the surface, it discarded the spent cocoon of the aeroshell. Slowed by atmospheric friction as well as by a giant parachute and a twenty-three-minute burn of its small descent rockets, the lander settled gently to the rusty ground and began to look around. Its twin cameras, capable of detecting organisms as small as a few millimeters in size, saw nothing that moved during that first close inspection.

But finding life at the macro scale was not Viking's mission. Coiled within the lander like the drive spring of a windup toy was an extensible arm of wire and steel. On the eighth Martian day, or sol, after the first landing, the arm unwound into the thin, cold atmosphere. Stretching some nine feet from the spacecraft, it lowered a scoop toward the Martian surface and began to dig. After scratching up a handful of soil, the arm retracted into its container and swung back toward the body of the craft, where it shook the soil into the cylindrical chimney of the biology instrument. Soil particles cascaded down into a kind of lazy Susan that allocated the material to the three experiments *(pages 22-23).*

The first teaspoonful of Mars went into the test chamber of Horowitz's laboratory, where it was inoculated with radioactive carbon dioxide. A xenon lamp was turned on to mimic Martian sunlight, and the sample was left to incubate. Another morsel was passed to Oyama's gas-exchange experiment, where the next day the test chamber containing the sample and Martian atmosphere was pressurized with carbon dioxide, helium, and krypton. Two drops of organic nutrient—so nourishing that Soffen dubbed it chicken soup—were added, and the chamber gas was analyzed at intervals for signs of microbial exhalations. A third sample, allocated to Levin's labeled-release experiment, received its first injection of about two drops of nutrients on sol ten and was then left alone so that its small inhabitants, if any, could metabolize their organic snack.

After filling the biology instrument on sol eight, Viking's sampling arm went back to the surface for another bite of Martian soil, which it was supposed to shake into the meshed top of the GCMS unit. This system had already shown a tendency to fail—an in-flight check had indicated that one of the three sample-heating ovens in the device was not working. Now, as the soil sample presumably rattled into the top of one of the ovens, the GCMS failed to flash a "level full" message to its masters on Earth. Another scoop was added, and still no message came. Reluctant to waste one of the two remaining GCMS ovens on a blank, mission engineers hurriedly began developing programs to pick up another sample on sol fourteen. But during this attempt, the sampler arm jammed as it retracted into its coil. Although the "anomaly," as the glitch was called, was quickly solved, Viking managers nervously pondered whether to take a chance that a sample had actually been dumped into the GCMS unit on sol eight, despite the lack of confirmation that it was full. Sol eight gleanings that had been put in the biology instrument were producing some bewildering results, and the Viking team thought the gas chromatograph/mass spectrometer might help clarify the situation if it could work on an equivalent batch.

On sol nine, Oyama's gas-exchange experiment had brought the first surprise. Analysis of gas in the test chamber showed fifteen times more oxygen than scientists expected, a value that eventually increased to more than 200 times the initial value over the next two sols. At first glance, it looked like a startling positive result in the search for microlife on Mars. But if this surge of oxygen came from Martian microbes, they were heavy breathers indeed compared to anything on Earth. A change of this magnitude, the team felt, almost had to be chemical, but it was ambiguous enough that no one was ruling anything out yet. On sol sixteen, the next part of the experiment began. More nutrients were added, and the sample began a 196-sol incubation period. During this time, no further trace of released oxygen was detected.

More puzzles were in store as the results of Levin's labeled-release experiment came in. The first injection of nutrients, on sol ten, produced a sudden surge of radioactive carbon dioxide in the chamber, a quick reaction very similar to that seen from organisms in terrestrial tests. If the gas came from

A Step-by-Step Inspection

To the world's space engineers, the safe touchdown of two identical Viking landers on Mars in 1976 was a momentous feat: Since 1971 four Soviet descent modules had either crashed or missed the planet entirely. Best of all, the delicate and highly sophisticated contents of the Viking twins survived the landings in perfect condition.

Tucked into the compact body of each lander were two oblong containers *(red circle, opposite)*, each about the size of a small microwave oven. The containers, capped with chimneylike structures to receive samples of the Martian soil, were stuffed with miniaturized versions of the equipment in a well-stocked biochemical lab. One held a set of three small life-detecting experiments *(below)*. The other was a scaled-down gas chromatograph/mass spectrometer *(page 24)*, a device designed to analyze material for so-called organic compounds, the carbon-based molecules typical of terrestrial life.

In each Viking lander, the hands of the human biochemist were replaced by a single clawlike scoop at the end of an extensible arm, shown on the opposite page. Made of two layers of spring steel that could retract into a coil or straighten into a rigid, ten-foot cylinder, the arm repeatedly reached out from the body of its lander to scratch up bits of Martian soil with the scoop. Once a soil sample was acquired, the arm retracted into its coil and swung left to shake the soil into one of the two minilabs. Analysis of the sample soon began and would continue for up to seven months, with all the findings relayed periodically to Earth via a circling Viking orbiter.

A Hardworking Package

Viking's thirty-four-pound biology instrument system was designed to perform a series of three major experiments—pyrolytic release, gas exchange, and labeled release—described on the next two pages. Among the 40,000 needed parts were tiny ovens, breakable ampules of nutrients, flasks of radioactive gas, Geiger counters, thousands of transistors, and a xenon arc lamp to simulate the Sun of Mars.

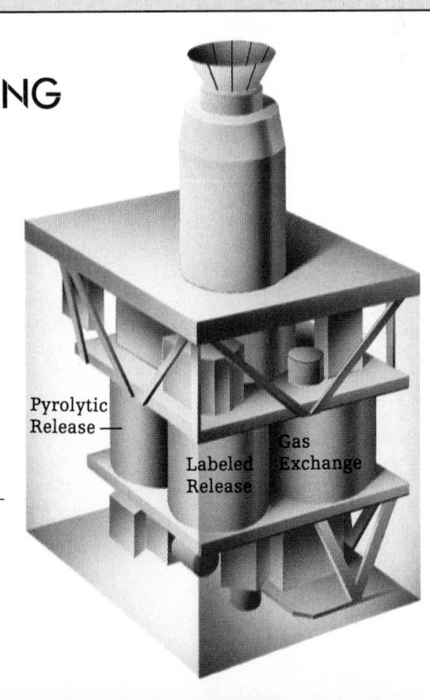

Pyrolytic Release: Is It Alive?

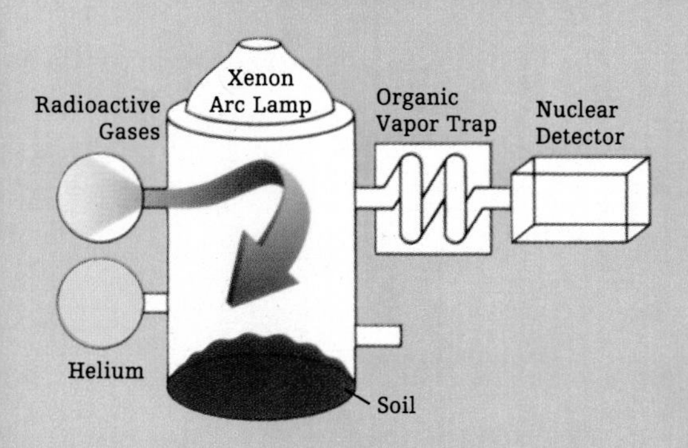

A Viking test cell containing a cubic centimeter of Martian soil (about one-fifth of a teaspoonful) fills with carbon dioxide and carbon monoxide *(green)*, gases known to be in the Martian air but here radioactively labeled with carbon 14.

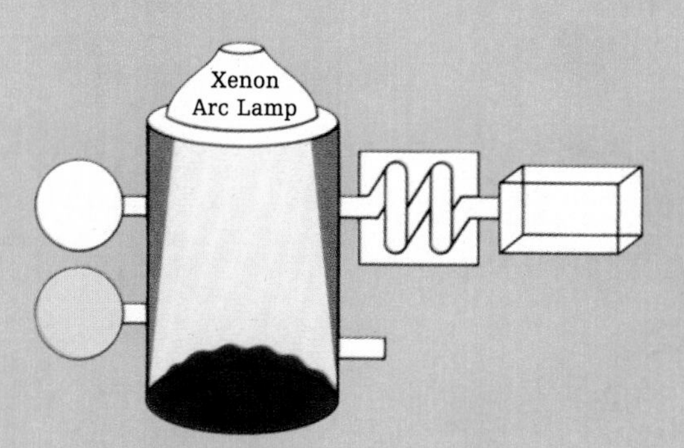

The test cell's xenon arc lamp turns on, simulating all but a small portion of the spectrum of Martian sunlight. After incubating for five days, any organisms in the soil sample will have assimilated some radioactive carbon from the gas.

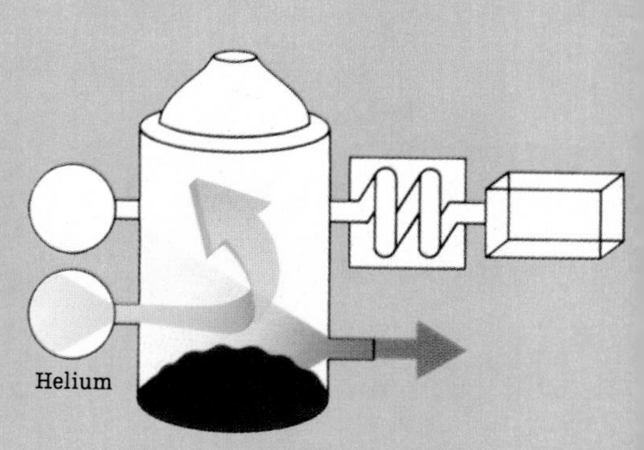

At the end of the incubation period, the light turns off, and helium gas *(yellow)* is released to flush out the atmosphere. The test cell then rotates to the pyrolysis head, joining with it to form a miniature oven.

Gas Exchange: Does It Breathe?

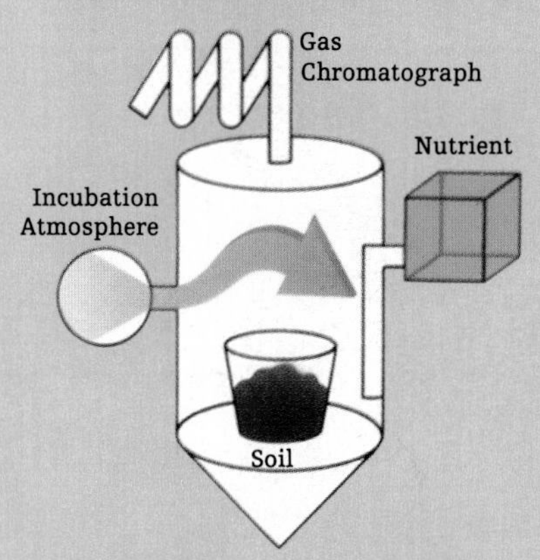

The second Viking test cell, with a cubic centimeter of soil in a small cup, fills with Martian atmosphere, supplemented by helium, and krypton *(arrow)*.

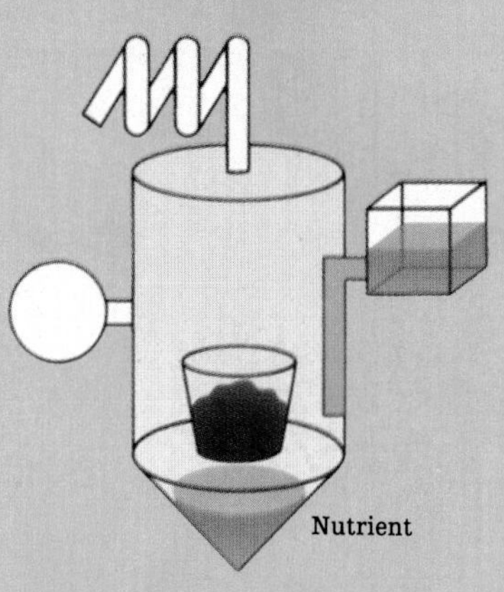

A water-based nutrient containing amino acids, vitamins, salts, and other chemicals trickles to the cell floor without coming in contact with the soil sample.

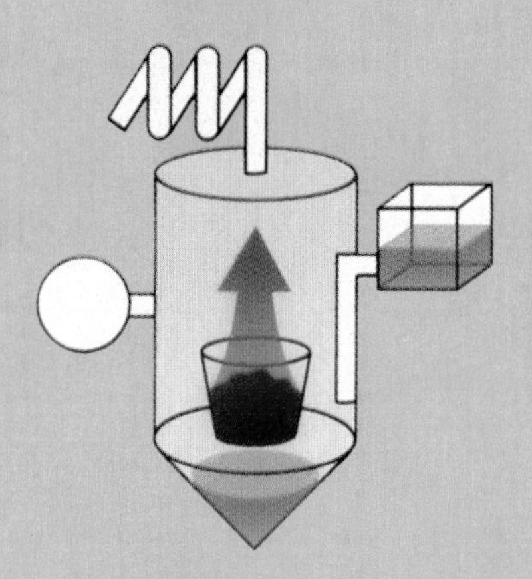

Although the liquid nutrient does not touch the soil sample directly, its partial evaporation creates a humid environment. The soil reacts with the water vapor, thereby releasing a gas *(arrow)*.

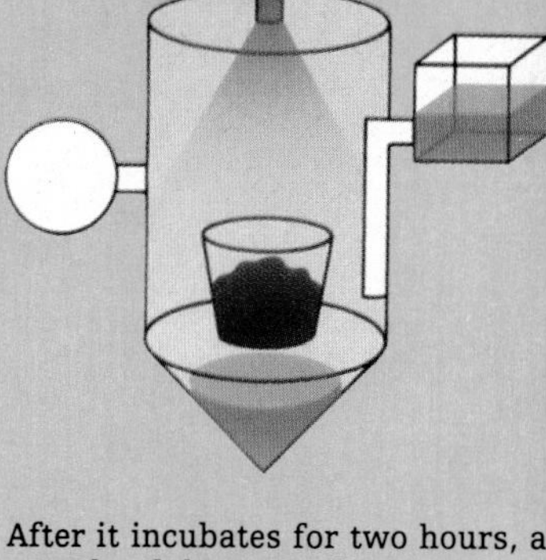

After it incubates for two hours, a sample of the test-cell atmosphere is chemically analyzed by a detector-equipped gas chromatograph *(top)*. Incubation continues, and the analysis is repeated at intervals for up to a week.

Labeled Release: Does It Eat?

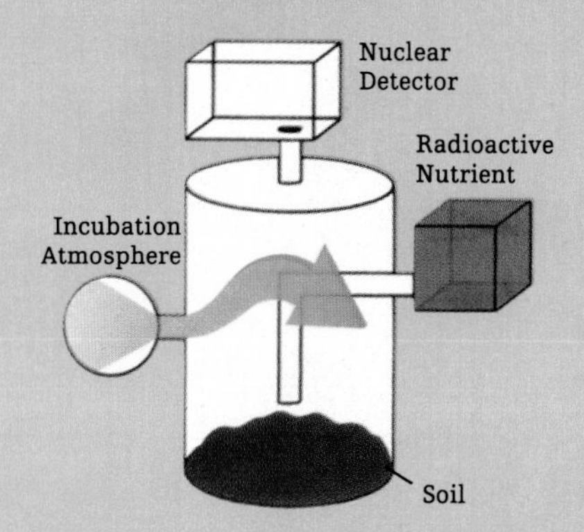

In the third experiment, a cell holding half a cubic centimeter of soil fills up with Martian atmosphere *(orange)* containing carbon dioxide, nitrogen, argon, and traces of other gases.

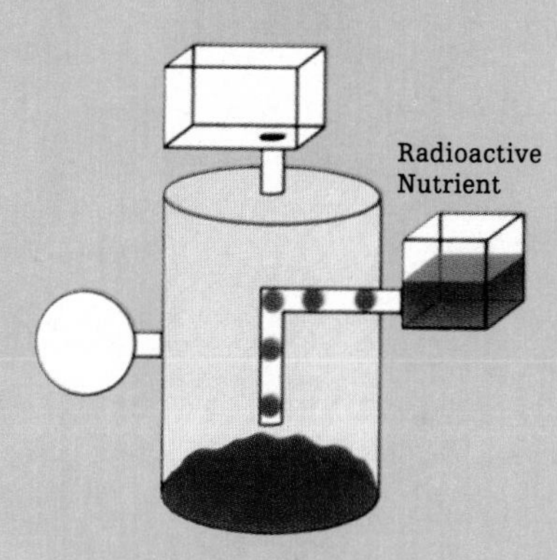

A snack of radioactively labeled, water-based nutrient *(green)* drips onto the sample, which then incubates at 41 to 80 degrees Fahrenheit for periods ranging from days to months.

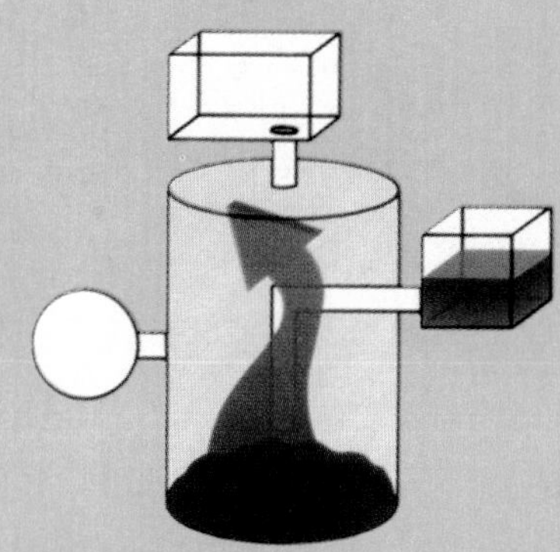

As the radioactive nutrient and soil interact, they give off gases *(dark green)* marked by radioactive carbon 14. The level of radioactivity suggests whether microorganisms are present.

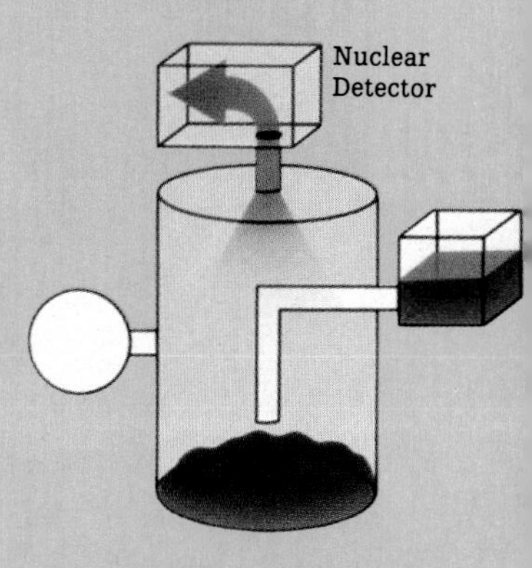

A nuclear detector above the cell continuously checks the amount of carbon 14 released into its atmosphere. Gradually increasing levels of carbon 14 might suggest the presence of microorganisms.

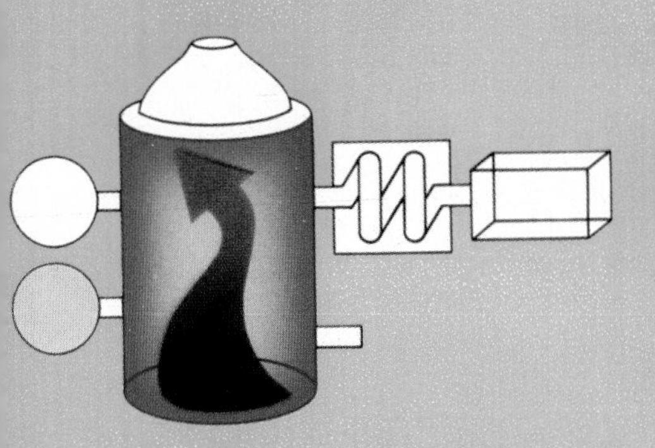

Next, the test cell heats to 1,157 degrees Fahrenheit. At this temperature, any organic material formed by Martian microbes should pyrolyze, or decompose, into volatile vapor *(arrow).*

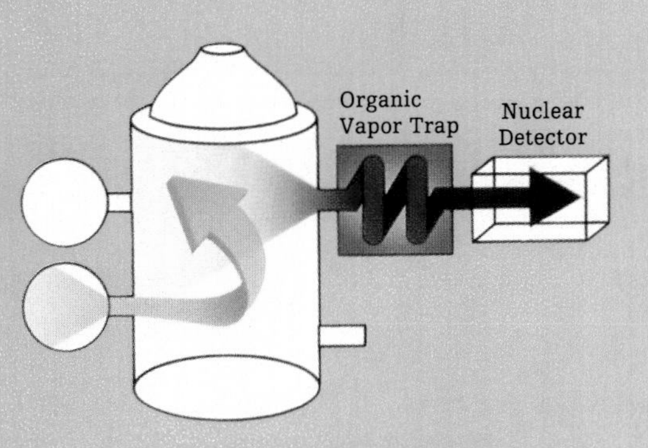

Helium *(yellow)* sweeps the volatized matter into an organic vapor trap. As the trap heats, the fragments react with its copper-oxide lining to form carbon dioxide, whose carbon 14 content can be measured by the nuclear detector at right.

RESULTS: Of the nine pyrolytic-release experiments on Mars—six by *Viking 1,* three by *Viking 2*—seven showed some carbon assimilation by the soil, a response resembling the activity of some Earth microbes. However, one of the seven trials had been performed on a heat-sterilized sample. For heat-treated soil to show signs of carbon assimilation suggested to most scientists that the results were a life-mimicking effect of the peculiar surface chemistry on Mars rather than evidence for Martian life.

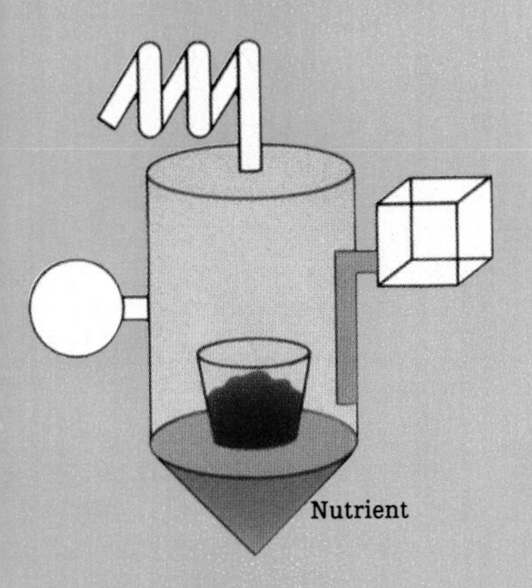

As the experiment continues, more liquid nutrient is added, this time deliberately wetting the soil sample. The contents of the cell are again left to incubate.

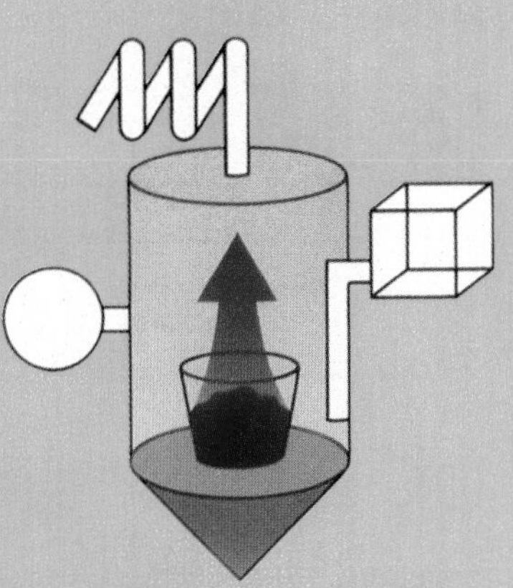

Any organisms in the soil should begin to feed on the nutrient, perhaps consuming Martian atmospheric gases as well, then release waste products such as carbon dioxide, methane, and other gases *(arrow)* into the cell.

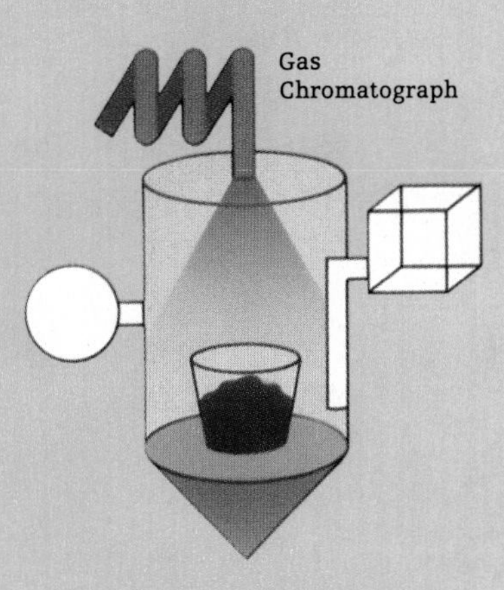

Over the ensuing six and a half months, the gaseous mixture forming in the test cell is repeatedly sampled and analyzed by the gas chromatograph.

RESULTS: In the first two and a half hours of the gas-exchange experiments, a surprising amount of oxygen showed up in the readings, far in excess of the oxygen expected in the Martian atmosphere. But after that abrupt jump, very little additional oxygen was produced, leading researchers to suspect that a chemical process, rather than plantlike, oxygen-exhaling life forms, had been at work. Later tests with sterilized soil confirmed their theory; the dead soil produced the same oxygen excess. As with the pyrolytic release *(above),* researchers attribute the results to chemical reactions with the Martian soil.

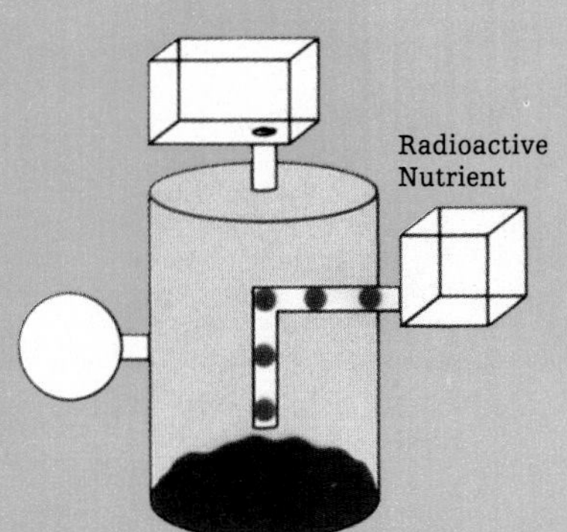

After about a week has elapsed, a second ration of nutrient *(green)* is added to the sample, which is otherwise unchanged.

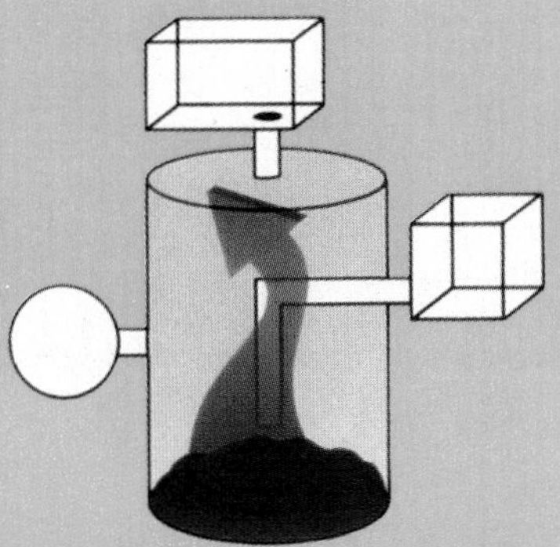

If the soil contains microorganisms, the addition of more nutrients should again produce a burst of radioactively labeled gas *(green arrow).*

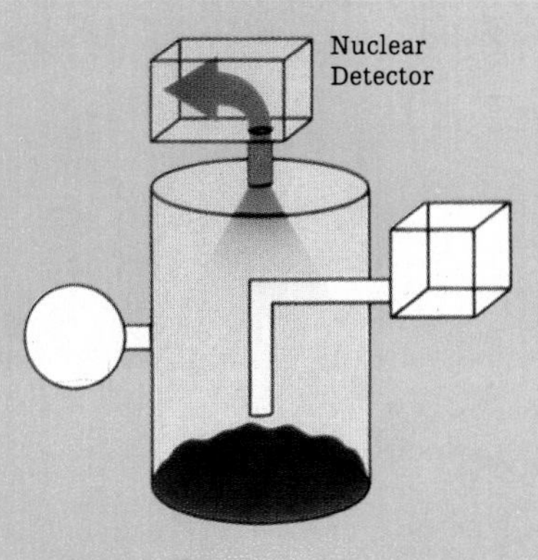

The nuclear detector measures carbon 14 gas *(green)* throughout the experiment. Increases could indicate microbial activity; low levels would suggest that either the first response was merely chemical or the organisms died.

RESULTS: Each lander conducted three labeled-release tests. In all, a sharp rise in carbon 14 followed the first dose of nutrients. When the soil was heat sterilized and the tests repeated, the initial carbon 14 release did not occur. In the absence of organic compounds *(overleaf),* most scientists theorized that the carbon 14 increases came from a reaction between the nutrient and heat-sensitive soil chemicals. But a few think that heating could have killed microorganisms in the soil and that the experiment did detect microbial life on Mars.

Hunting for Organic Chemicals

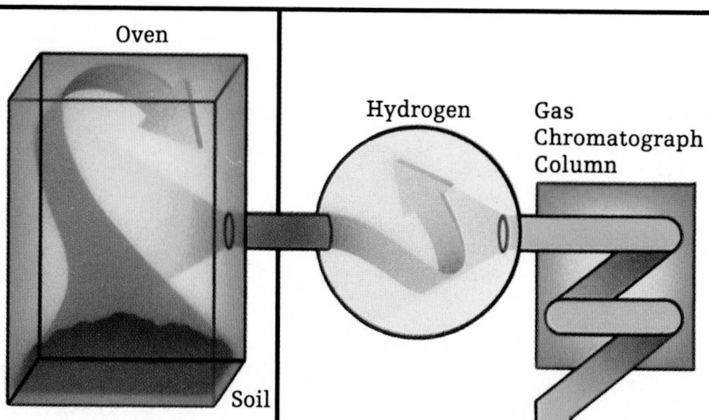

1 Martian soil is ground up and put into a small oven to begin its journey through the gas chromatograph/mass spectrometer, shown below. Heated to 932 degrees Fahrenheit, the soil is reduced to gas, which is transported onward by a hydrogen flow system *(yellow).*

2 Hydrogen carrier gas and volatized soil chemicals flow together into the coiled gas chromatograph column, a heated filtering system of glass beads coated with an oily polymer. The more complex a gas component is, the slower its trip through the column will be.

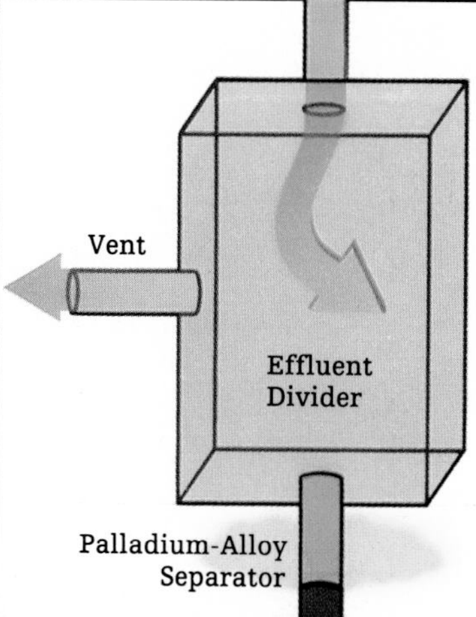

3 Each filtered gas progresses in turn to the effluent divider, a device that controls pressure by venting excess gases *(left).* Beyond the divider, the gas travels through a palladium alloy separator that permits hydrogen *(yellow)* to escape but passes on the original gas *(brown).*

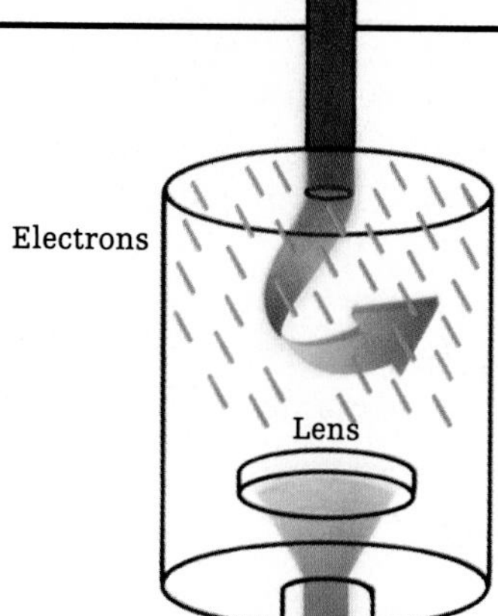

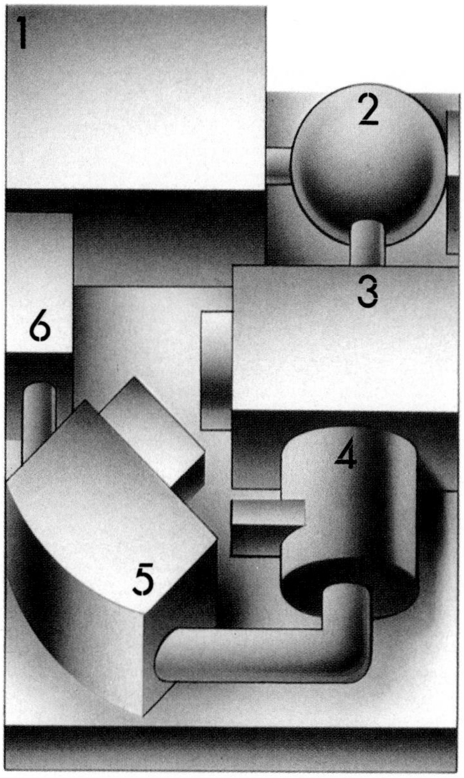

RESULTS. Before the Viking experiments, scientists assumed the Martian surface would be rich in carbon-based organic compounds even if there was no Martian life: Incoming meteorites should have deposited organic chemicals on the planet. But the gas chromatograph/mass spectrometers, instruments sensitive to within a few parts per billion, detected no trace of organic compounds—suggesting that the chances for contemporary Martian life are extremely remote.

4 At the entrance to the mass spectrometer section, high-voltage current bombards incoming gas, ionizing, or electrifying, its molecules by knocking away electrons. Magnets and a lens then focus the ionized gas *(purple)* into a thin stream, which flows on to the magnetic chamber.

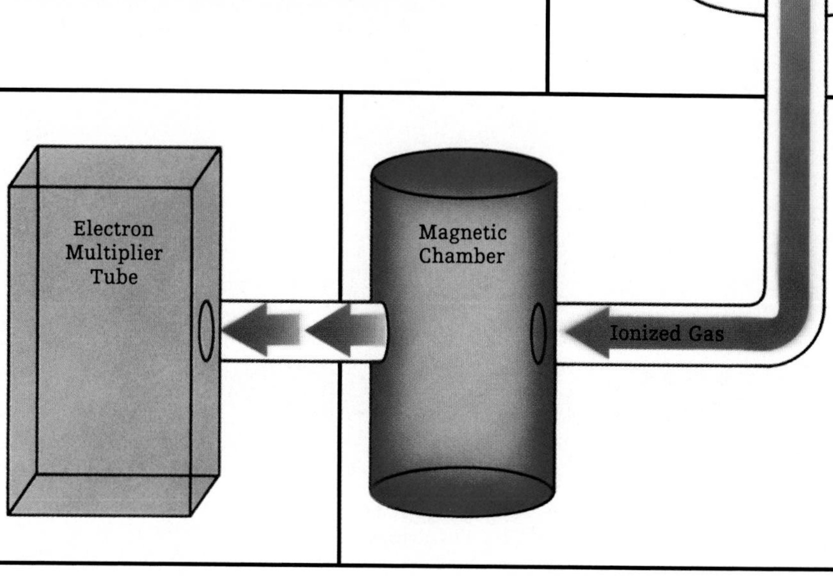

6 Finally, the ionized, presorted gases *(purple)* stream into an electron multiplier tube that identifies the organic fragments present. The chemical analysis is translated into an electrical signal, which will later be transmitted to Earth.

5 As the ionized gas passes through a magnetic field *(blue),* its chemical constituents are sorted by their molecular weights to reveal the nature of any organic compounds in the original soil sample.

Martian microbes, a second injection should produce a repeat performance. On sol eighteen, another sip of nutrient broth was injected into the sample, with results that echoed the gas-exchange experiment: Nothing happened.

Although two of the three experiments seemed to be indicating that Mars was lifeless, the third had moderately more encouraging news. On August 7, sol eighteen, Horowitz told the Viking press corps that his pyrolytic-release experiment had detected a reaction, adding "there's a possibility that this is biological." The sample had taken up, or fixed, a surprisingly large amount of radioactive carbon. While small compared to the rate of carbon fixing in the life-rich soils of Earth, the Martian result was well above the background levels the Viking team had previously established with sterile soil samples.

To verify that the effect was indeed biological and not a chemical illusion, a second sample was heated to 347 degrees Fahrenheit. Any life adapted to the cold Martian climate would presumably be killed off, and the next analysis of gas in the test chamber would show a complete absence of fixed carbon. As it turned out, the drop in the amount of fixed carbon was significant, down by some 88 percent. Once again, this was an ambiguous finding: Had the decrease been 100 percent, the source would almost certainly have been biological. But was the leftover 12 percent of fixed carbon sufficient to declare the source of the reaction purely chemical—or could it have come from a feeble spark of Martian life?

TO THE COURT OF APPEALS

In the absence of a clear verdict, as one NASA historian put it, the gas chromatograph/mass spectrometer became Viking's court of appeals. On August 6, sol seventeen, the team led by Klaus Biemann, a forty-nine-year-old Vienna-born scientist whom colleagues called "a renaissance man in mass spectrometry," took the plunge. Gambling that the GCMS had indeed been loaded with a sample on sol eight, the team instructed the device to proceed with its analysis of the soil.

The next day, Biemann reported that the instrument had worked well. The first run of the GCMS returned 300 mass spectra, measurements of charged particles arranged sequentially by mass. Biemann's team scrutinized this first sample for organic compounds—carbon-based molecules containing hydrogen, nitrogen, and oxygen—and for evidence of water. They found little sign of water in this set of data and no organic compounds at the instrument's parts-per-billion limit of sensitivity. Six days later, they had the instrument reheat the same sample to 932 degrees Fahrenheit, in the hope that higher temperatures would break down organics into material the instrument could detect. But, again, the device found no organics.

The second Viking lander arrived at Utopia Planitia on September 3, sol forty-five, and conducted similar experiments with similar results. Klaus Biemann spelled out the implications: The surface of Mars contained even less organic material than the most inhospitable patches on Earth, the near-sterile soils of Antarctic dry valleys. Not only were such compounds evidently not being produced on the planet's surface, but any organics imported by meteorites, an occurrence common on Earth *(page 43-45)*, were apparently being destroyed in the Martian environment. "If we use terrestrial analogies," Biemann summed up after the mission, "we always find that a large amount of organic material accompanies living things—a hundred times, thousand times, ten thousand times more organic materials than the cells themselves represent." Since neither one of the Viking probes had found signs of that organic waste, Biemann simply could not make a case for the existence of microorganisms at the two test sites.

The story did not end there. Mission scientists and others continued to analyze Viking's data over the next decade. Eventually they came up with scenarios to explain how the planet's chemical makeup may have produced imitations of biological reactions. "Just as a living organism can decompose a steak by eating it and digesting it," biology team leader Harold Klein told NASA historians after the mission, "the steak can also be decomposed by being thrown into sulfuric acid, with roughly the same end results."

Some hypotheses link the explanation to the planet's color. Because the thin Martian atmosphere does not block the Sun's ultraviolet radiation, these rays oxidize, or rust, the iron that makes up 13 percent of the planet's surface soil, producing Mars's characteristic hue. The ultraviolet bath, the theory goes, would also break organic compounds into inorganic chemical debris. Further incoming radiation would dissociate carbon dioxide, which makes up 95 percent of the Martian atmosphere, to form such highly reactive products as atomic hydrogen, hydroxyls, and atomic oxygen. These in turn would interact with the iron-rich soil to produce so-called superoxides, another destroyer of organic materials. In much the way hydrogen peroxide, catalyzed by iron compounds in a wound, releases oxygen bubbles, superoxides could have produced the oxygen seen in both the gas-exchange and labeled-release experiments. Similarly, oxidation of the nutrient compounds by superoxides could have created the illusion that carbon was being assimilated by organic matter, as seen in the pyrolytic-release experiment.

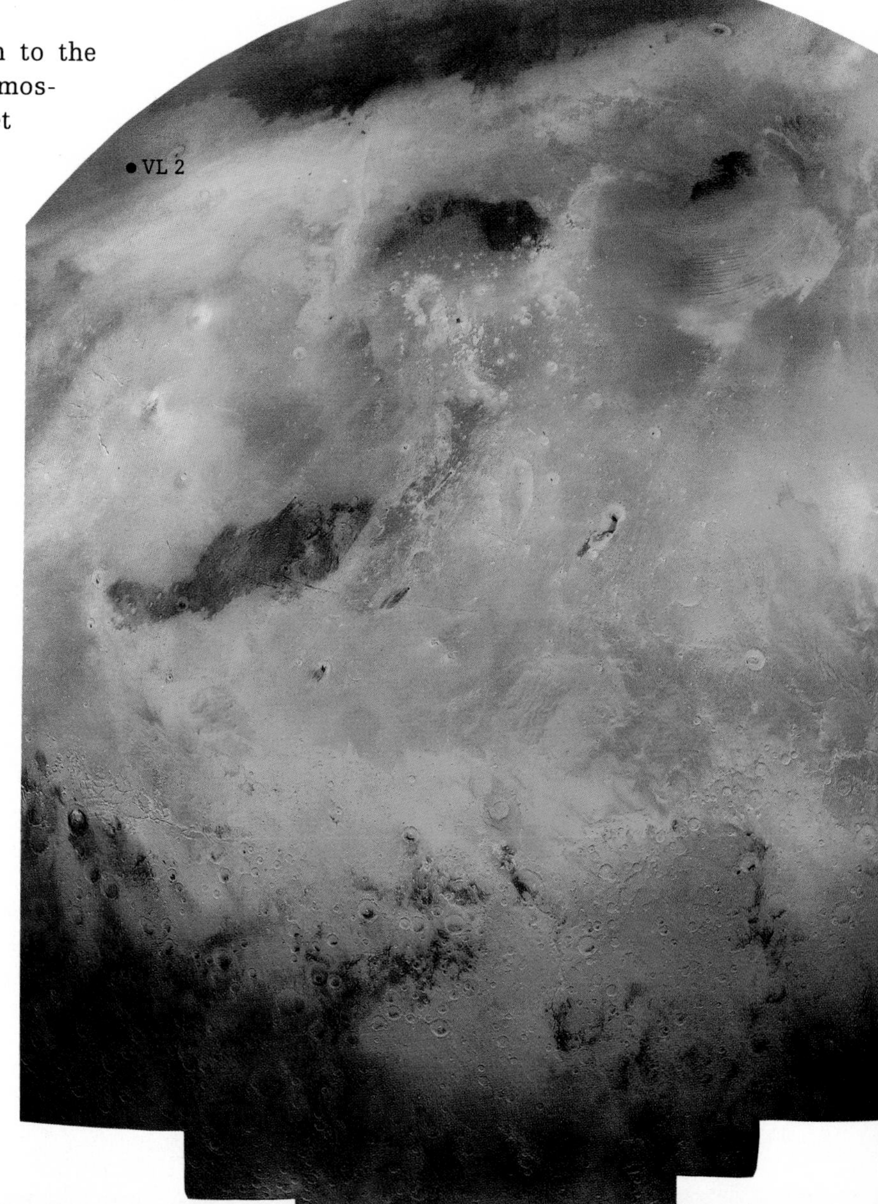

According to the theory such processes, and the efficient distribution of superoxides across the planet by frequent dust storms, made it likely that no organic matter exists anywhere on the Martian surface. Between ultraviolet radiation from the Sun and the natural oxidation of any organics in the soil, Mars offered a lethal one-two punch to any recognizable life forms. The rust-covered planet was not merely a hostile environment—it appeared to be self-sterilizing. In this vast red world, richly marked by old signs of planetary vitality and ancient floods, seemingly so Earth-like that human imagination had populated it with a race of noble beings, nothing lived.

This was dismaying for exobiologists. Finding no direct signs of life had been tolerable; the most they had hoped for in any case had been a sparse colony of microbes. The lack of even indirect evidence—not a trace of the organic material with which terrestrial life litters its surroundings—was something else again, "a real disappointment," as Gerald Soffen put it.

When the two Viking landers dropped to the Martian surface in 1976, they arrived at sites half a world apart and in very different climatic zones. As indicated on the two photomosaics shown here, *Viking 1 (right)* touched down in the Martian tropics at 22.5 degrees north latitude, *Viking 2 (left)* nearly 178 degrees of longitude to the west and a temperate 47.5 degrees above the equator. The mosaics, assembled from hundreds of images taken by the Viking orbiters circling 20,000 miles overhead, show the thin clouds and water-worked terrain that still tantalize exobiologists.

A HAUNTING VISION

But Martian life, having survived for centuries in the human imagination, was not going to be exterminated so easily. Even Klaus Biemann hedged his pronouncement by noting that Viking's results did not necessarily preclude some other kind of living mechanism, one that "would protect its organic constituents very well, and, therefore, avoid this waste of a scarce commodity." For example, he suggested, Martian microbes might be buried deeper than Viking probed. Other scientists speculated that Mars was dotted with oases that might be little gardens of life. Harold Klein noted, "We can certainly say that it is not rampant, but we can't be sure there isn't some scraggly form of life for which we just haven't found the right nutrients or the right location or the right incubation temperature or the right environment."

Of the members of the Viking biology team, only one refused to rule out the possibility that the results obtained on Mars had been biological and not an illusion of Martian soil chemistry. Speaking a dozen years after his labeled-release experiment had been performed by Viking, biochemist Gilbert Levin bristled at being marked as a man who claimed to have found life on Mars. "What I *have* claimed," he said, "is strong evidence for life on Mars," adding that "it is more probable than not that we discovered life."

Though clearly in the minority, Levin based his statements on the results obtained by his labeled-release experiment before, during, and after the Viking mission. Before the mission, he said, tests of the flight-type model of his instrument in a chamber used to simulate conditions on Mars produced results with "plain old California soil" that were initially close to the results later obtained on Mars. The difference was that on Mars the initial signal, a sharp increase in the carbon dioxide levels in the test chamber, was not followed by another increase when more nutrient was added eight sols later. Levin countered that he had obtained almost identical results with Antarctic soil—known to be nearly, but not entirely, sterile—in which the first carbon dioxide peak was followed by no further increases. He also noted that gas chromatograph/mass spectrometer tests on Antarctic soil had failed to detect organics that were discovered later when the soil was subjected to so-called "normal wet lab" analysis. Thus, according to Levin, Viking's GCMS, while capable of detecting organics down to the parts-per-billion level, was not sensitive enough to detect the much smaller concentrations he still believes are in the Martian soil.

Perhaps the most intriguing of Levin's results, however, was obtained not on Mars but at his Maryland laboratory in 1981, when a colleague brought Levin a rock from Antarctica. Sliced open, the rock revealed fine bands of dark material composed of endolithic organisms—microbial creatures that live inside rocks. "They scraped the stuff out," Levin recalled, "and put it into the labeled-release experiment." The response was "almost identical to what we got on Mars." Although many years have passed since the Viking mission, Levin remains haunted by its ambiguities. "Not a day goes by," he said in 1988, "that I don't think of that experiment."

BUILDING BLOCKS OF LIFE

For a long time, the planet was all harsh disorder, enduring torments from within and without. Quakes and volcanoes savaged its surface. Comets and meteors hurtled down through yellow skies. Seas boomed against naked rock. Like the rest of the half-billion-year-old Solar System, the young Earth was sterile.

It would not remain so for long. In that turbulent era, the shower of particles and larger objects from space brought an infusion of carbon compounds, and more such molecules were created by the interaction of the atmosphere with lightning and solar energy. Somehow, the chill seas and cold, wet shores of Earth stirred the chemicals into life. In another two and a half to three billion years, that life would fashion complex cells with central nuclei *(above)*. Multicellular plants and animals would follow in less than a billion years.

Why Earth? Time has veiled the answer. Scientists can only analyze the makeup of present-day life—illustrated in part on the following pages—and speculate on its probable origins. As they do so, they cannot help but ponder a related mystery: whether similar biologies have evolved anywhere else in the galaxy, on far-off planets steadily orbiting other suns.

Carbon, Earth's Vital Atom

At the atomic scale *(below)*, earthly life consists almost entirely of a sparse handful of elements—hydrogen, nitrogen, oxygen, phosphorus, sulfur, and most important, carbon. Because carbon easily forms multiple bonds with other atoms, it acts as a kind of glue, cementing together the pieces of life's complex molecules. Scientists consider it the likeliest basis for life anywhere, a hypothesis that led NASA to include carbon-compound detectors aboard the Mars Viking landers. Even the language of chemistry honors carbon's biological role: Compounds containing carbon and its companion hydrogen are termed organic, all others, inorganic.

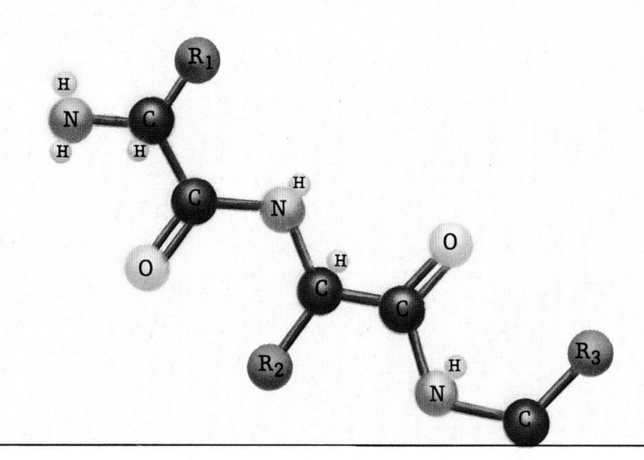

Why Carbon Combines

Carbon atoms are among the most versatile known, forming chemical attachments, or bonds, with as many as four other atoms at once. A single carbon atom can thus hold together a formaldehyde molecule *(below)*, and a row of atoms dotted with carbon may serve as backbone for a protein (*background, right*).

The reason carbon bonds so easily is that it has relatively few electrons. In a carbon atom, as in any other, electrons orbit a proton-neutron core in what may be thought of as concentric shells. In all atoms, each shell may hold a certain number of electrons. The inner shell accommodates as many as two, the next one, eight. But a carbon atom has only six electrons; with two electrons in the inner shell and four in the next, four vacancies in the outer shell remain.

Carbon atoms tend to fill the gaps with electrons from other atoms nearby. A carbon atom may share an electron belonging to each of four other atoms, a process that creates four distinct single bonds. Or, a carbon atom may fill two or three of its vacancies with electrons from one atom, forming a double or triple bond. In the case of formaldehyde, a carbon atom forms one double bond and two singles—two electrons come from an oxygen atom, one electron each from two hydrogens.

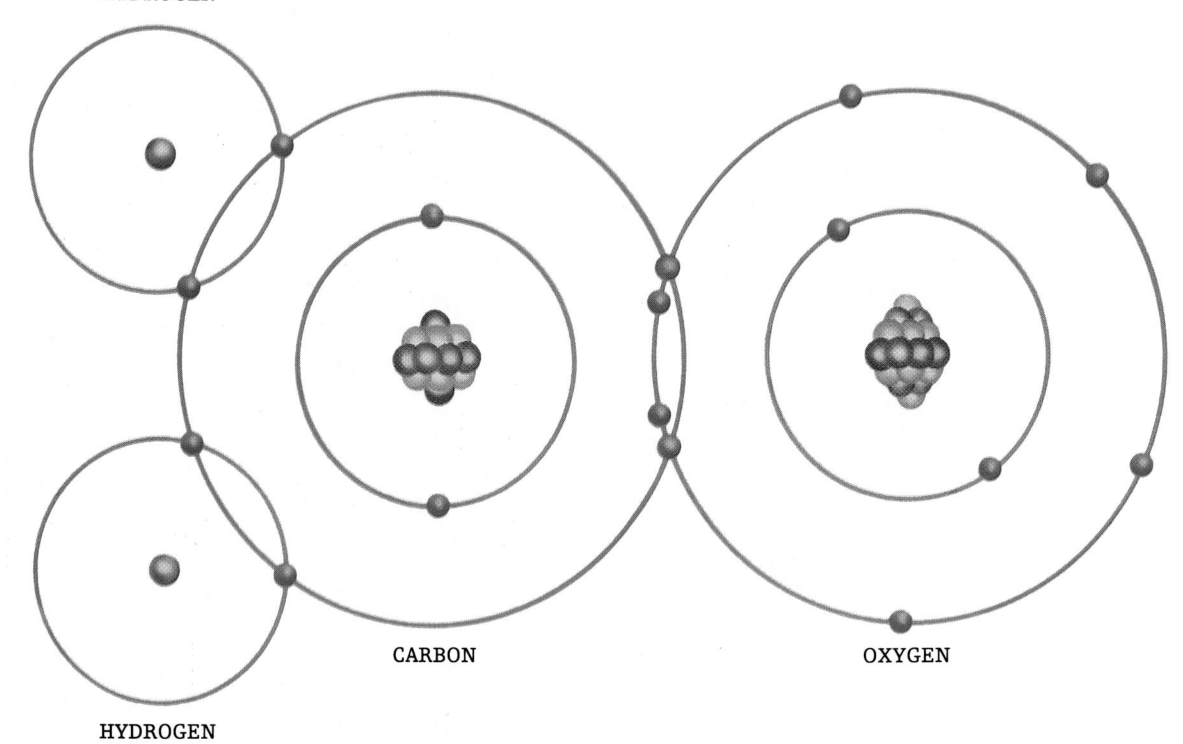

Many organic molecules are relatively simple, from four-atom formaldehyde *(box)* to the compounds called amino acids, a family of structurally similar molecules that are made up of about ten to thirty atoms apiece. Found in abundance on Earth, small organics are still more common in space, where they have been detected in dust clouds, on comets and meteorites, even floating in the atmosphere of Jupiter.

But complex, self-regulating systems that live, reproduce, and die require far more sophisticated molecules, some encompassing more than 10,000 atoms. Crafted in the course of millions of years of prebiotic chemical evolution, these are the organic polymers—giant chains, rings, lattices, and globules built from chemical units known as monomers, of which amino acids are one variety. Organic polymers fall into four classes: carbohydrates, fats and oils, nucleic acids like DNA *(pages 32-33)*, and commonest of all, proteins *(pages 36-37)*. Each protein is a tangle of one or more peptide chains like the strand below. Peptides consist in turn of hundreds of amino acids, bonded in a sequence that varies from protein to protein. Theoretically, hundreds of different kinds of amino acids are possible. Only a score are found in living things on Earth, and carbon life on other worlds may well employ an entirely different amino acid set.

Marked by multiatom side chains, which chemists term radicals *(silver Rs)*, two amino acids draw together, expelling a water molecule in the process *(right)*. The water forms as the left-hand amino acid sheds a hydrogen and an oxygen atom, and the other unloads a single hydrogen. With those atoms gone, other atoms—in this case, nitrogen and carbon—are free to bond the two amino acids together. As similar reactions occur over and over, hundreds of amino acids link up to form a peptide chain *(below)*, one of the strands making up a protein.

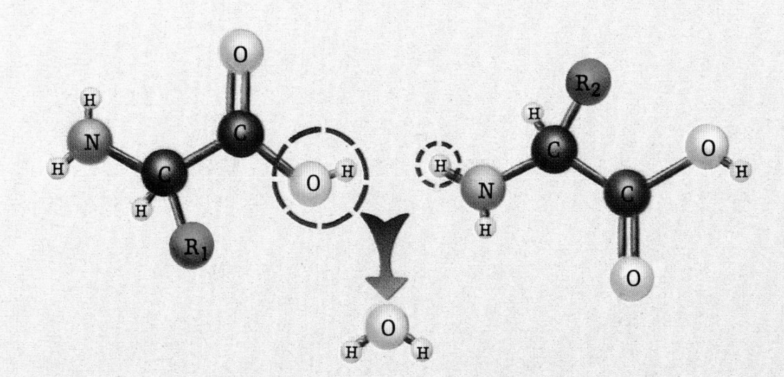

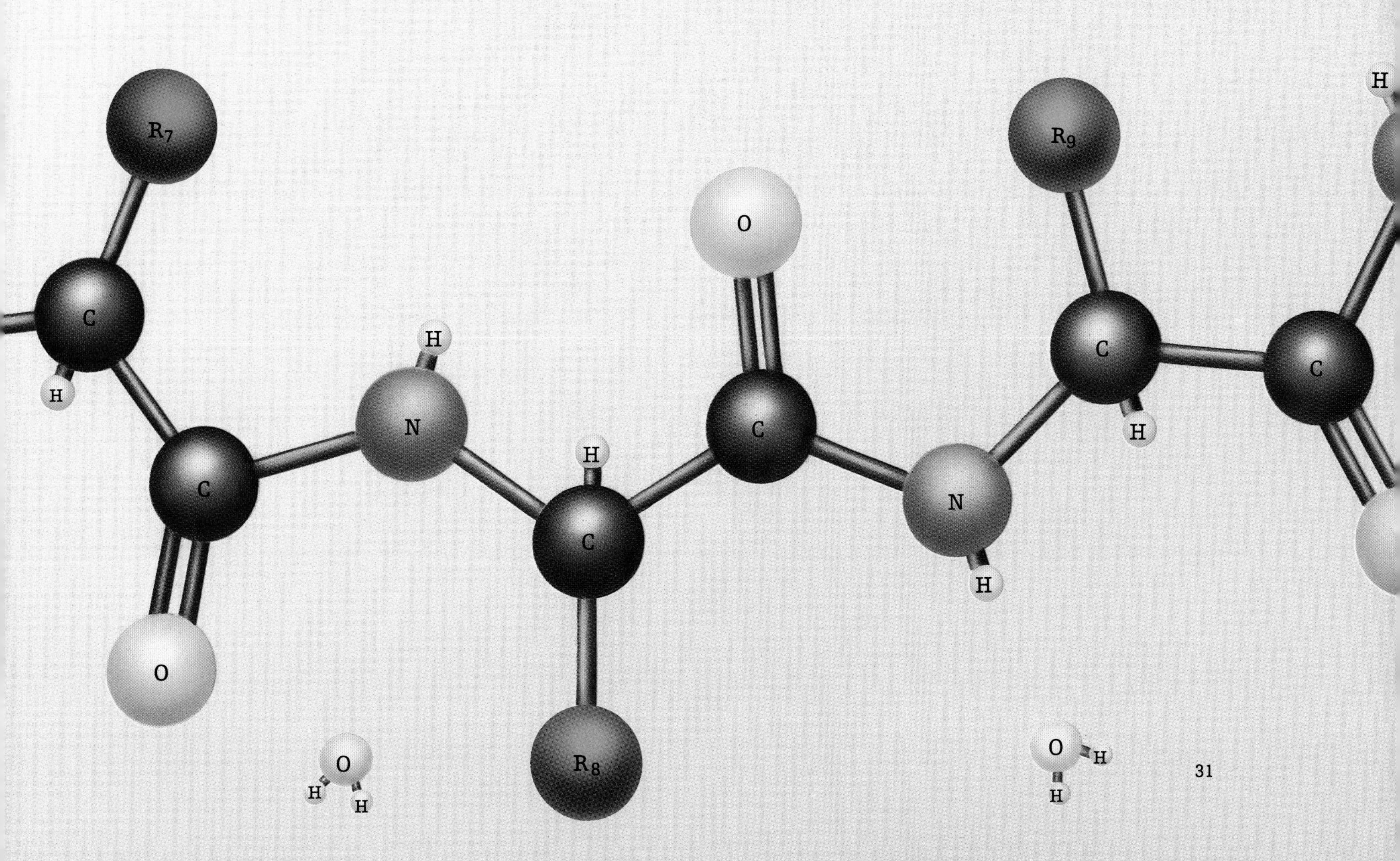

A Molecular Code

The essence of life on Earth—and presumably elsewhere as well—is the ability to grow and reproduce. Both activities require stored plans for a creature's particular set of proteins and instructions for putting them together. Terrestrial life accomplishes these tasks by means of a molecule called DNA, short for deoxyribonucleic acid.

Like a protein's peptide chain *(pages 30-31)*, a molecule of DNA consists of simple chemical units linked systematically together. In proteins, the units are amino acids; in DNA and other nucleic acids, they are compounds called nucleotides.

A nucleotide contains a phosphate, a sugar, and a section called a base, of which there are four kinds. The bases known as cytosine and thymine each incorporate one hexagonal carbon-based ring; the other

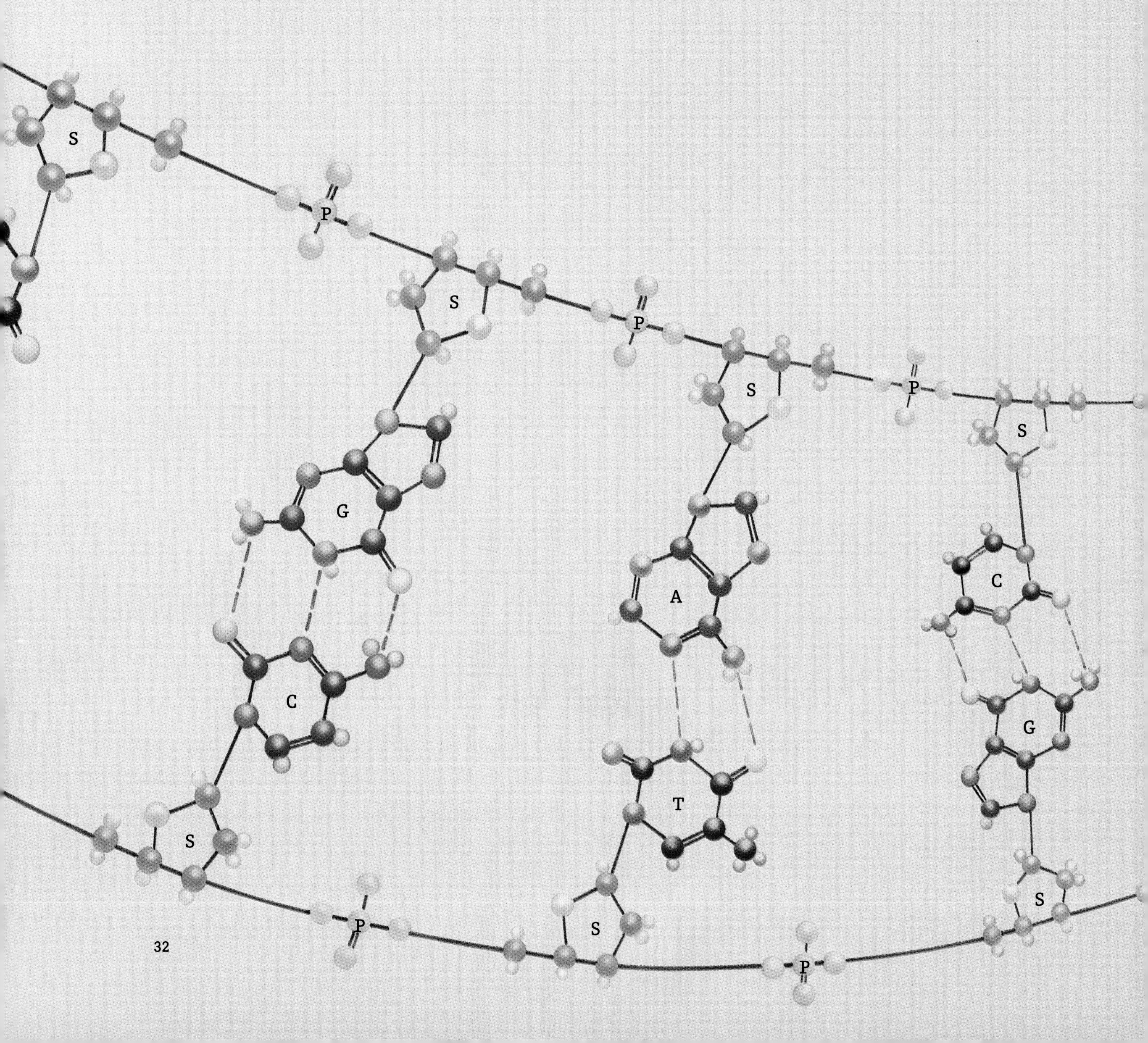

two bases, adenine and guanine, have two rings each.

The nucleotides in a DNA molecule form two parallel strands, linked together by their bases and twisted like a spiral staircase. The ordering of the nucleotides is critical. Like letters of a biochemical alphabet, nucleotides are meaningless in isolation; only in sequence do they spell out lengthy instructions, as shown on pages 34-35.

For electrochemical reasons, the way the bases bond to one another follows a set pattern. In a DNA molecule, adenine always bonds to thymine and guanine to cytosine. As a result, the sequence of nucleotides along each strand is a chemical mirror image of its mate. Because this allows for a kind of cross-check of the stored facts, the error rate during replication is less than one in a billion. Exobiologists assume that any form of complex extraterrestrial life would require a similar mechanism for replication.

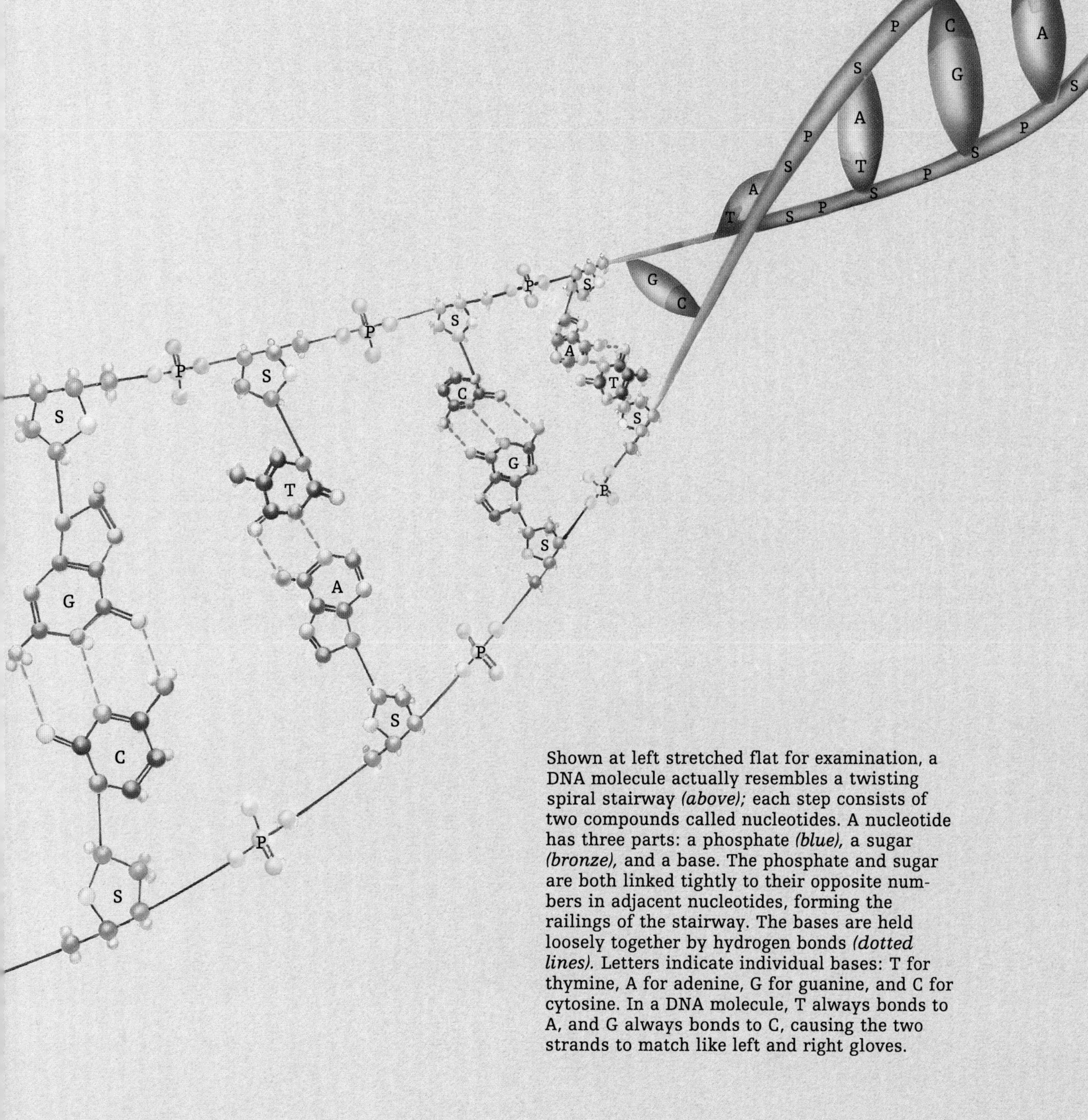

Shown at left stretched flat for examination, a DNA molecule actually resembles a twisting spiral stairway *(above)*; each step consists of two compounds called nucleotides. A nucleotide has three parts: a phosphate *(blue)*, a sugar *(bronze)*, and a base. The phosphate and sugar are both linked tightly to their opposite numbers in adjacent nucleotides, forming the railings of the stairway. The bases are held loosely together by hydrogen bonds *(dotted lines)*. Letters indicate individual bases: T for thymine, A for adenine, G for guanine, and C for cytosine. In a DNA molecule, T always bonds to A, and G always bonds to C, causing the two strands to match like left and right gloves.

Reading the Genetic Message

Like a biological librarian, DNA preserves the information needed to fashion the protein molecules that are the basis of life. A kindred compound called RNA, short for ribonucleic acid, helps turn those instructions into reality. Although no one can be sure how or when DNA and RNA first emerged on Earth, their appearance marked a turning point in the chemical evolution toward terrestrial life—a long, crucial step from the random jostlings of amino acids and nucleotides in an organic slush.

The key to the DNA-RNA partnership is a shared language, spelled out along the DNA strands in three-letter "words" called codons. A codon is made up of the bases of three successive DNA nucleotides. With

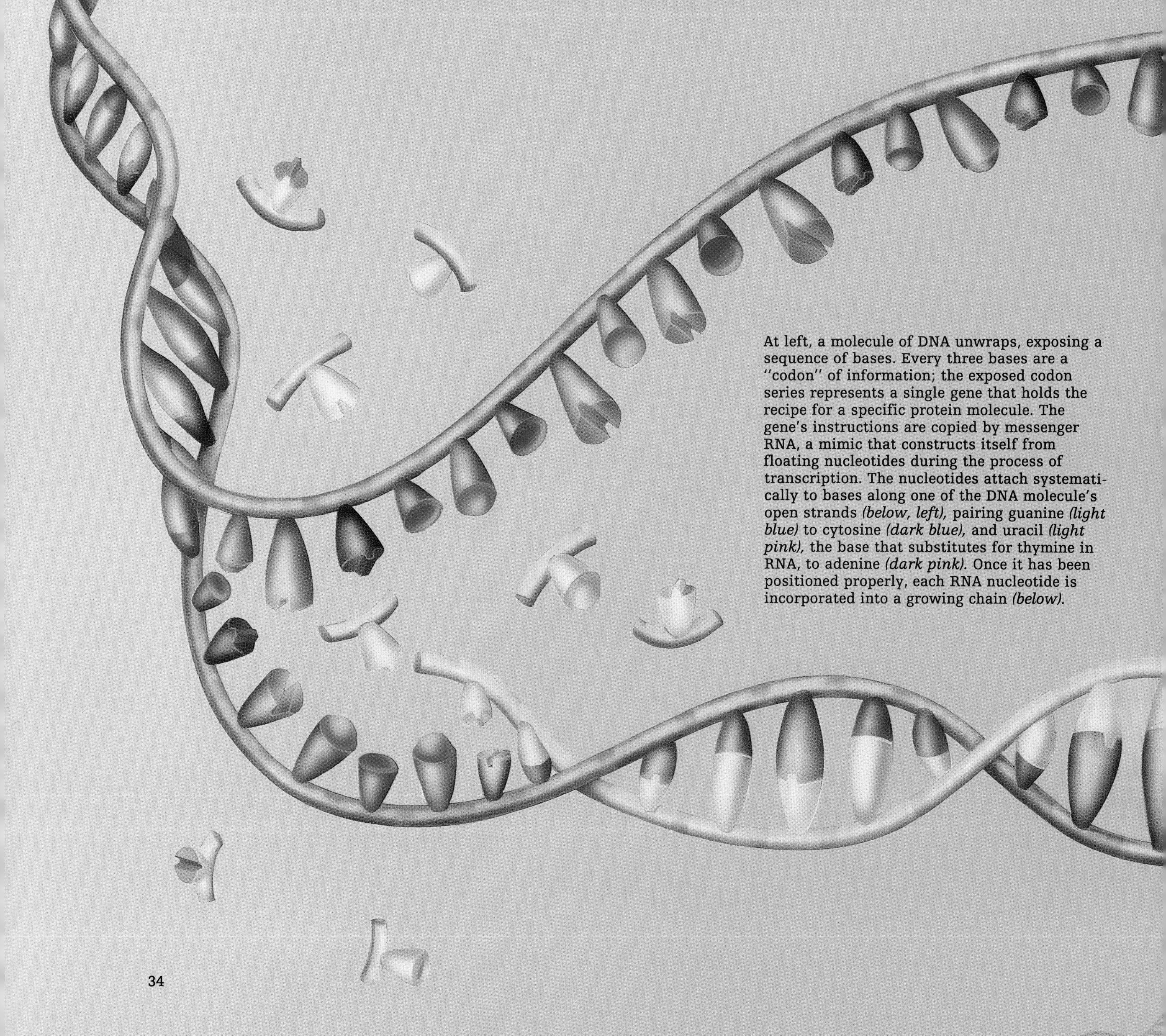

At left, a molecule of DNA unwraps, exposing a sequence of bases. Every three bases are a "codon" of information; the exposed codon series represents a single gene that holds the recipe for a specific protein molecule. The gene's instructions are copied by messenger RNA, a mimic that constructs itself from floating nucleotides during the process of transcription. The nucleotides attach systematically to bases along one of the DNA molecule's open strands *(below, left)*, pairing guanine *(light blue)* to cytosine *(dark blue)*, and uracil *(light pink)*, the base that substitutes for thymine in RNA, to adenine *(dark pink)*. Once it has been positioned properly, each RNA nucleotide is incorporated into a growing chain *(below)*.

four different bases, there are sixty-four possible three-base sequences, and thus there are sixty-four codons. The commonest codons simply specify a particular amino acid. For example, the combination of two guanines and a cytosine forms the codon for an amino acid called glycine.

If codons are words, genes are the sentences they form, beginning with a special initiator codon and ending with a terminator. In between a gene's message consists of a list of amino acids, arranged in the order required to make a particular protein. DNA's genetic messages are readily duplicated by messenger RNA, a molecule that effectively assembles itself during the copying process *(below)*. Incorporating DNA's instructions in its own structure, the messenger RNA then voyages out to the machinery of the outer cell, where it begins the manufacture of a specific protein molecule by following the recipe it carries *(pages 36-37)*.

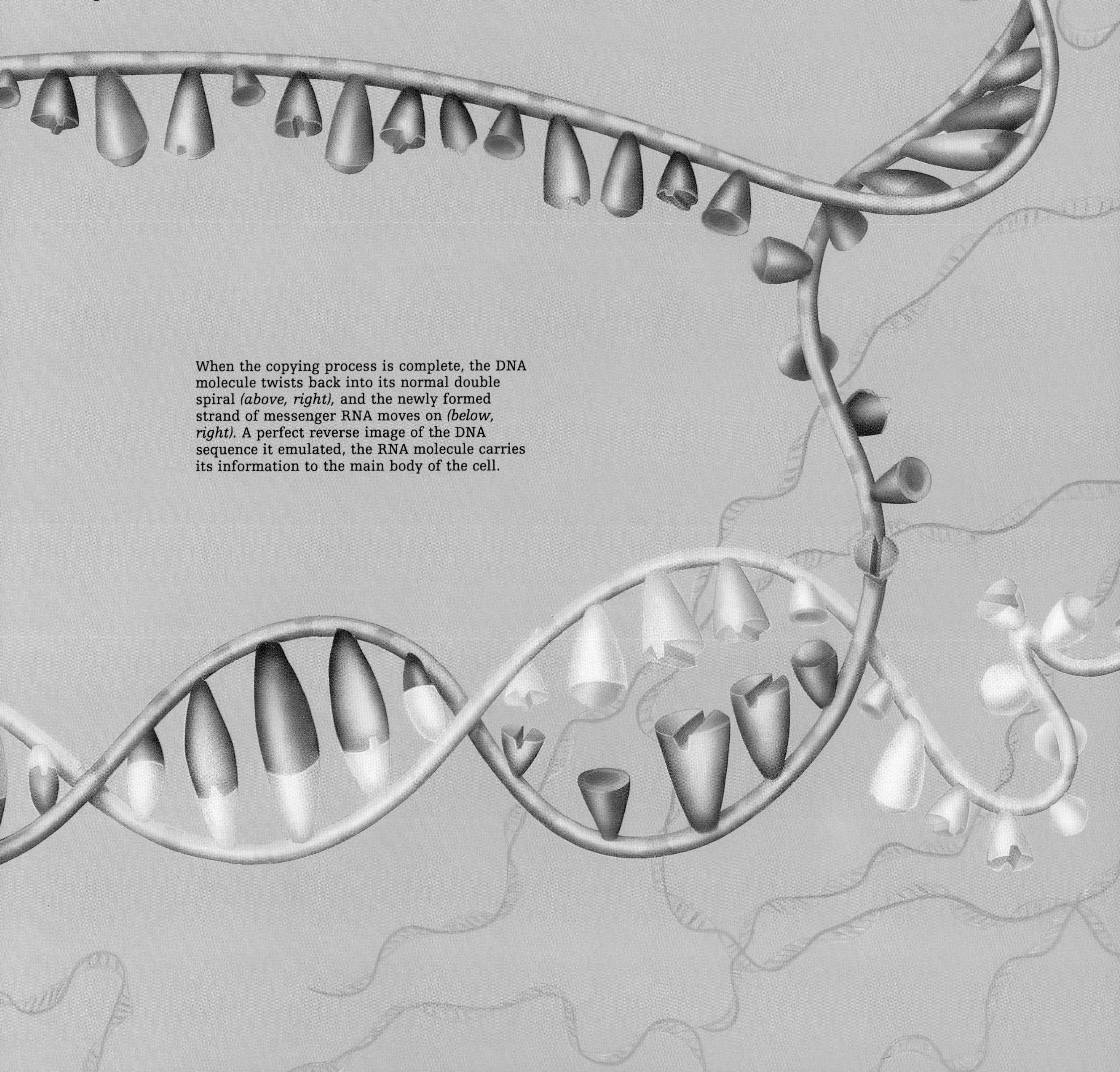

When the copying process is complete, the DNA molecule twists back into its normal double spiral *(above, right)*, and the newly formed strand of messenger RNA moves on *(below, right)*. A perfect reverse image of the DNA sequence it emulated, the RNA molecule carries its information to the main body of the cell.

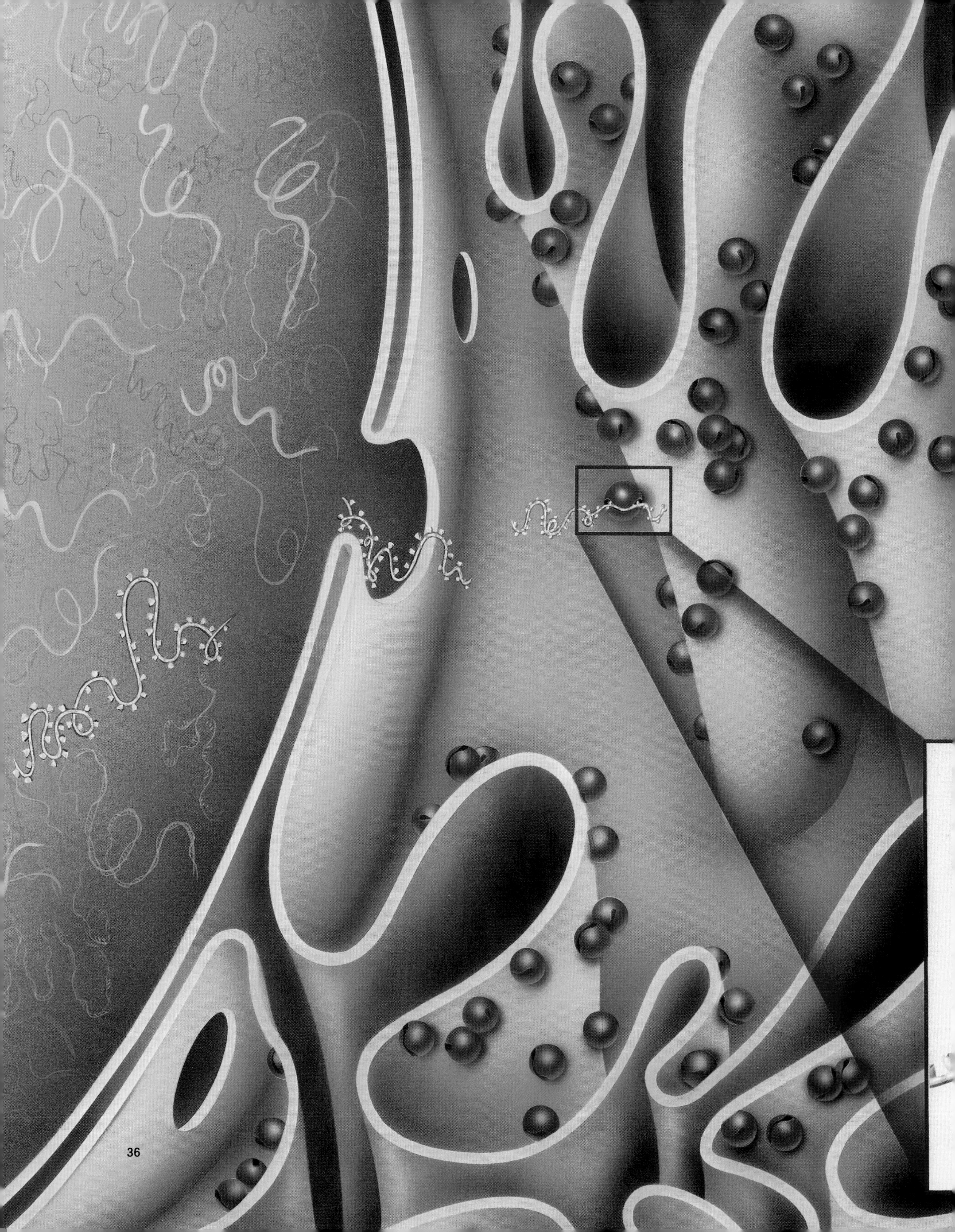

THE PROTEIN ASSEMBLY LINE

To translate genetic information into proteins, living things on Earth follow a complex manufacturing system. The illustration at left shows how a cell of a typical Earth life form goes about this task.

Work begins as a strand of messenger RNA *(pages 34-35)* enters the cell's protein-assembly area, carrying the genetic code for a particular protein. The messenger RNA wends its way through the watery interior of the cell in search of a structure called a ribosome *(brown)*. Typically a millionth of an inch across, these sophisticated protein-assembly machines reside on a mass of folded membranes called the endoplasmic reticulum. Ribosomes are equipped both to read the messenger RNA's orders and to carry them out.

Once the messenger RNA docks at a given ribosome *(below)*, the ribosome looks for the beginning of the RNA message, then attaches there. Messenger RNA proceeds to wriggle through the ribosome, allowing it to read the RNA codons in sequence. For each codon, the ribosome chemically signals for a specific variety of transfer RNA, a type of RNA whose job it is to deliver single amino acids. When the transfer RNA arrives, it touches down just long enough to unload its amino acid, which the ribosome links to a growing peptide chain *(below)*. The process is remarkably efficient; even in a bacterium, one ribosome can attach twenty separate amino acids to a peptide chain every second.

After the final codon has been read and its message obeyed, the ribosome releases a finished peptide chain into the cell. The peptide's own electrochemical properties will quickly wrap it and other peptides into the folded arrangement that forms a particular protein molecule. This molecule's work will depend on its identity: The protein known as collagen provides structural support in bone and ligaments, for example, while proteins called antibodies fight disease.

1 A single transfer RNA *(tan)* carrying an amino acid *(maroon)* nears the protein-assembly area of a ribosome *(brown)*. Not yet involved in the protein making, the transfer RNA floats in its cell's watery cytoplasm.

2 The ribosome decodes a strand of messenger RNA and signals for bits of transfer RNA that carry the specified amino acids. Here, the responding transfer RNA is about to connect to the messenger strand of RNA.

3 Once linked to the messenger RNA, the transfer RNA releases its amino acid to the ribosome, which fixes it to a peptide chain. The chain will soon fold and intertwine with others to create a functional protein molecule.

4 Once they are rid of their amino acids, the transfer RNA carriers drift back into the cell's fluid interior. There they will acquire some new amino acids and wait in readiness for construction of another protein to begin.

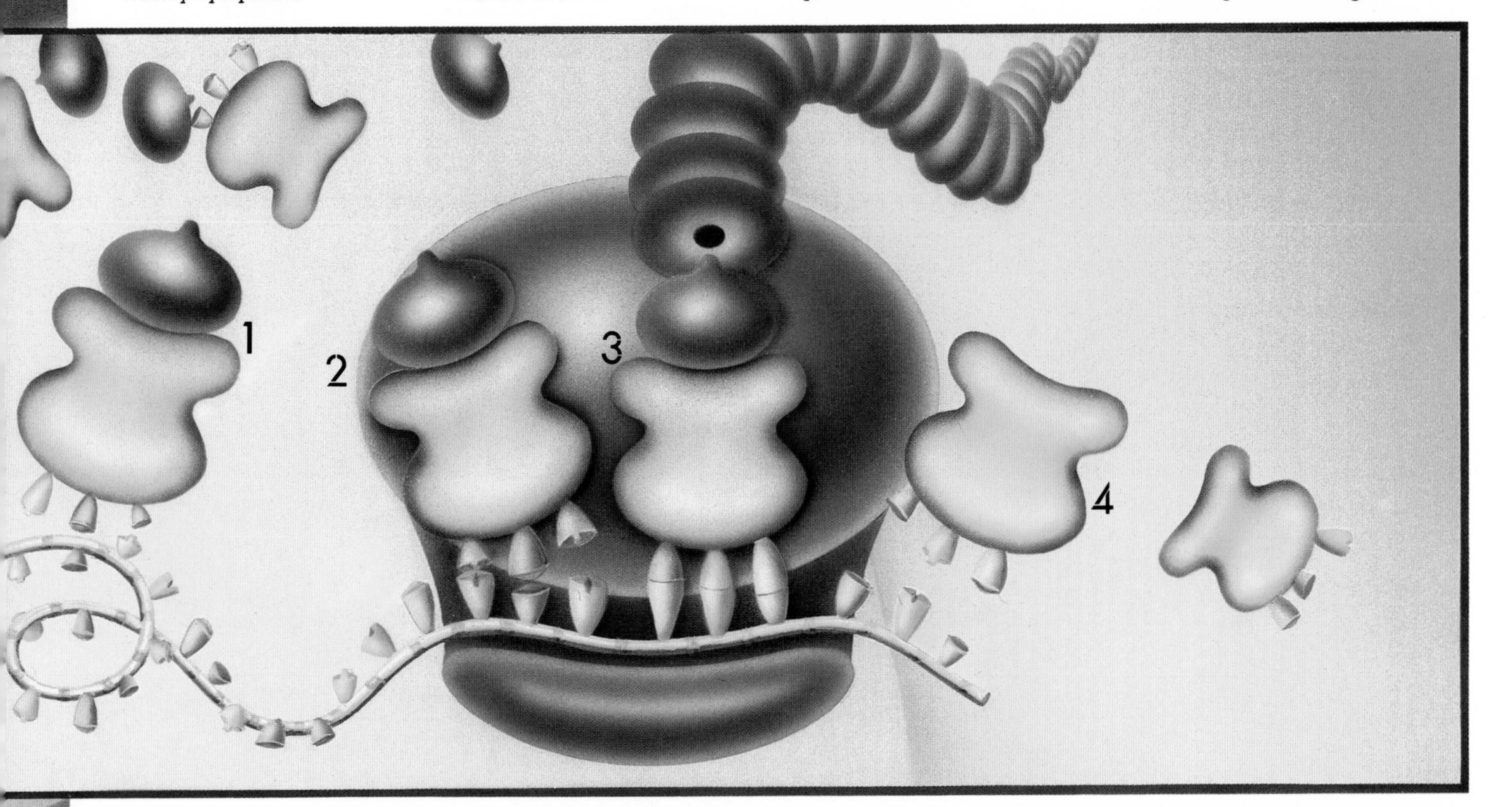

A MICROSCOPIC COMPLEX

The smallest unit of self-sufficient life on Earth is the cell, a tiny construct often no larger than a thousandth of an inch across. The first cells were simply a membrane holding in watery cytoplasm, within which DNA, RNA, and other organic compounds floated. Such cells persist today in common, single-celled bacteria called procaryotes, from a Greek phrase meaning "before the nucleus." Some procaryotes evolved an inner unit called the nucleus to hold and protect DNA. In such eucaryotes (from the Greek for "true nucleus"), the separate, outer cell region is filled with tiny mechanisms called organelles *(right).* Organelles allow eucaryotes to specialize and at times to combine and create elaborate, multicelled organisms.

The eucaryote cells of complex life forms are themselves highly complex. Within the nucleus are DNA's genetic codes as well as a still-mysterious unit called the nucleolus. Outside the nucleus, the ribosomes assemble proteins. Ribosomes generally perch atop the endoplasmic reticulum, whose sheets, sacs, and tubes weave through the cell's interior, carrying newly manufactured proteins, which are wrapped and dispatched by organelles called Golgi apparatuses.

The cell is fueled by the mitochondria, which convert nutrients such as sugars into energy, then store it in compounds called adenosine triphosphates, or ATPs, for later release. In green plant cells like that at right, additional organelles called chloroplasts help convert light into similar ATP packets. Attending to the cell's defense are so-called lysosomes, which are always wrapped in protective membranes; if ever uncovered, these chemical warriors would digest the cell itself. Finally, the whole cell is clad in a permeable membrane, sturdy enough to hold the cell cytoplasm but thin enough to allow nutrients in and fats, proteins, and waste out.

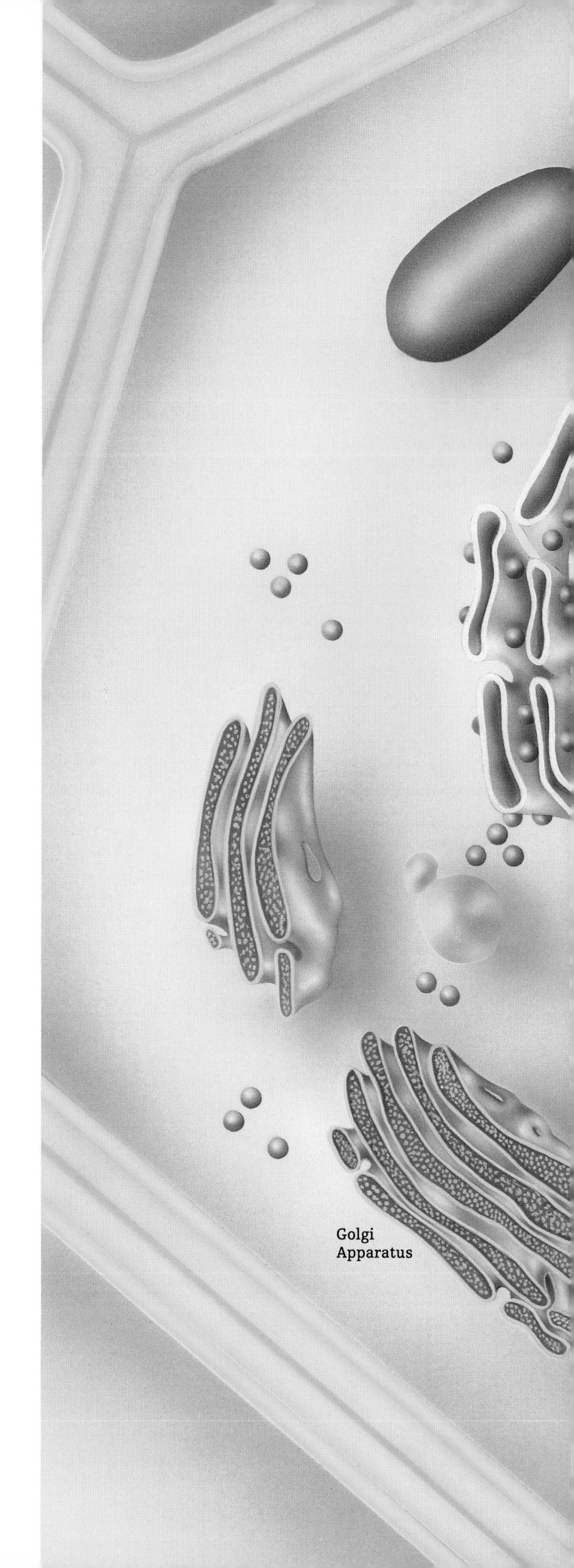

Shown here in cross section, a typical plant cell on Earth combines a central nucleus for information storage with an outer region that is full of tiny specialists, from bacteria-fighting lysosomes to photosynthetic chloroplasts.

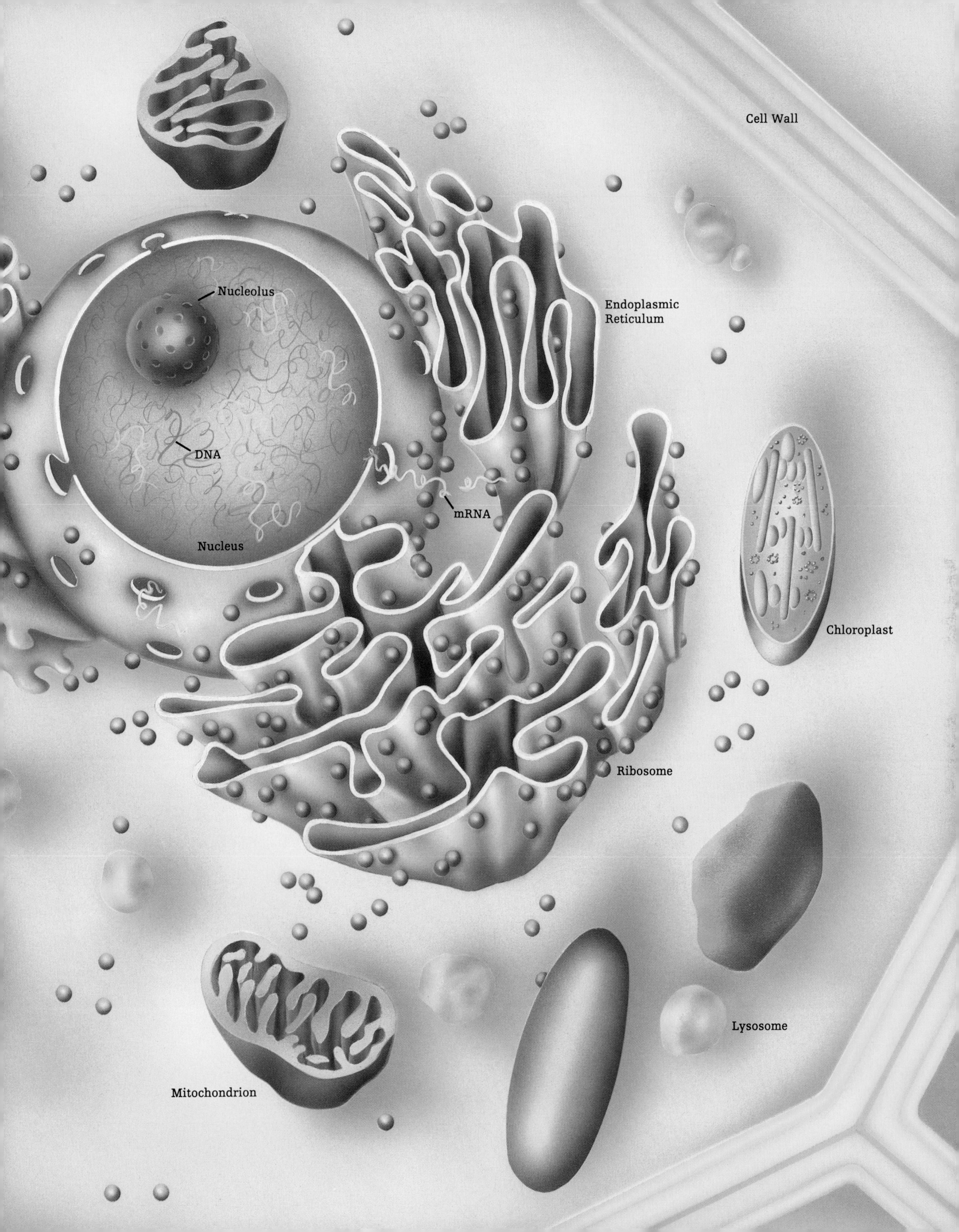

Cell Wall
Nucleolus
Endoplasmic
Reticulum
DNA
mRNA
Nucleus
Chloroplast
Ribosome
Lysosome
Mitochondrion

2/The Vital Spark

life, organizing clusters of organic molecules and, as crystals grew, duplicating new patterns in a primitive form of reproduction.

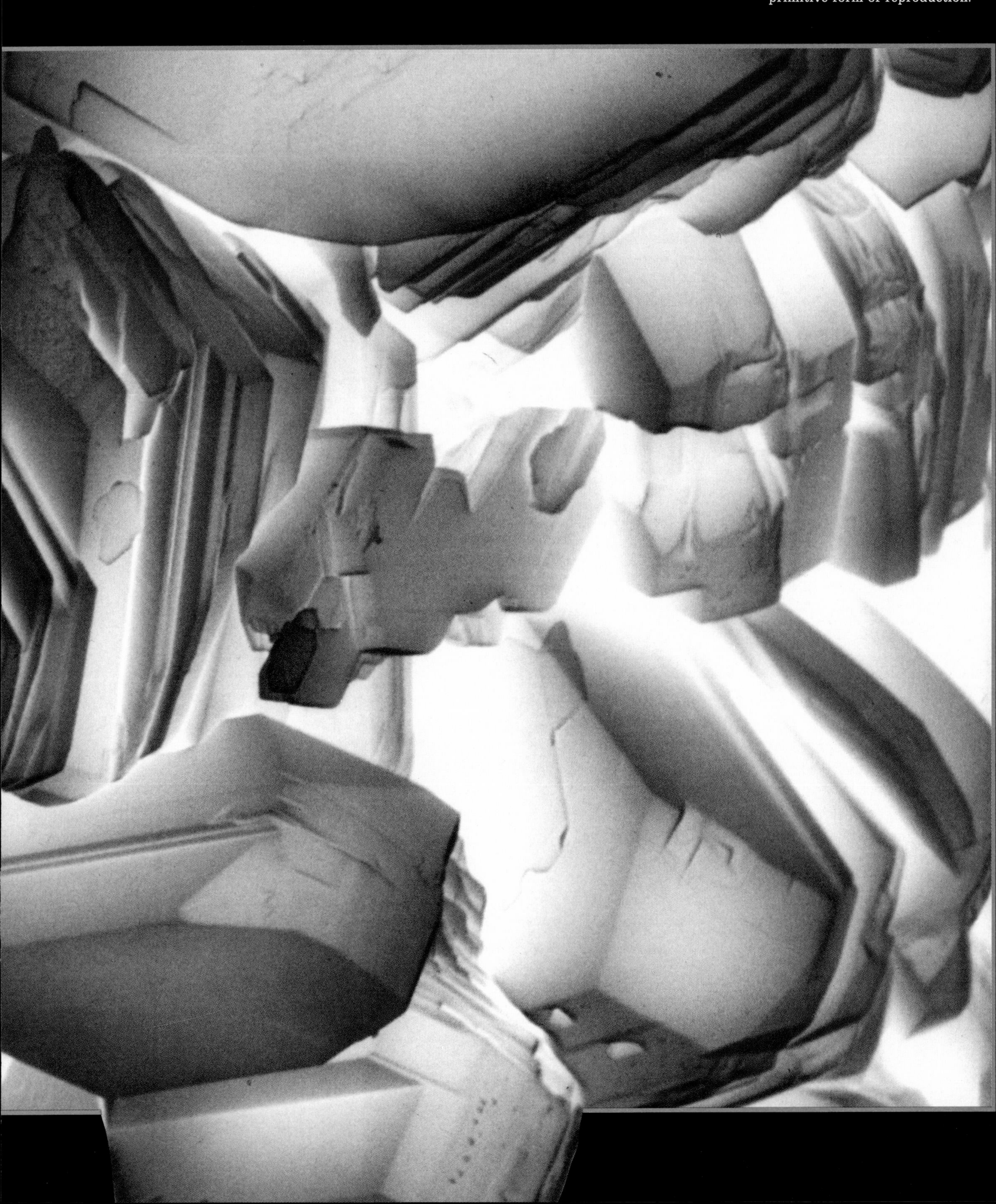

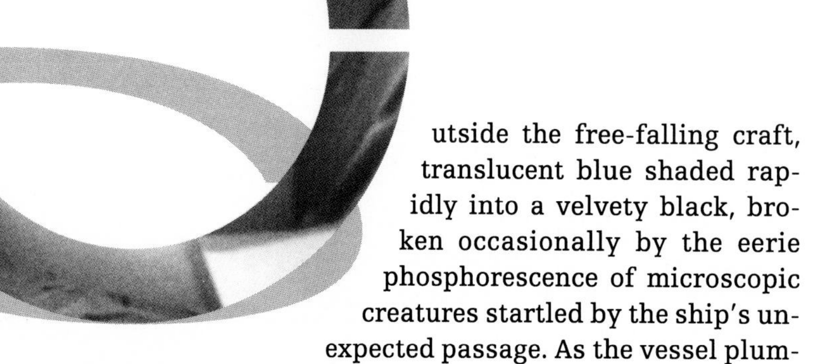

utside the free-falling craft, translucent blue shaded rapidly into a velvety black, broken occasionally by the eerie phosphorescence of microscopic creatures startled by the ship's unexpected passage. As the vessel plummeted, dropping a hundred feet per minute toward a zone of perpetual night, its three crew members nervously watched the ghost light flickering in the portholes and wondered if the cockpit's assorted echo sounders, sonar, and radar were keeping them on track. Peter Rona, a geophysicist from the National Oceanic and Atmospheric Administration, listened for the ping of the signals every ten seconds, an audio lifeline linking them with the mother ship overhead.

The near-freezing fluid closed around them like a great fist, squeezing the vessel's titanium alloy skin with pressures of some 5,000 pounds per square inch. Shivering, Rona and colleague John Edmond, a Massachusetts Institute of Technology geochemist, scanned the instrument panel for reassurance. When the sonar indicated they were just 300 feet above their hovering point, pilot James Hardiman adjusted the craft's ballast to arrest their fall and switched on the exterior flood lamps. Gray, eyeless creatures leaped into view, patches on their backs flashing silver. Then, as quickly as they had appeared, they were gone; the lights played over an empty desert of mud and rocks rising to a giant mound.

Hardiman steered toward the summit and finally brought the twenty-five-foot-long craft to a halt. At approximately 11:00 a.m. on May 23, 1986, *Alvin,* as the little ship was called, had completed a 12,000-foot dive to the volcanically restless edges of the Mid-Atlantic Ridge. Along this giant crack, in a hell of crushing pressure and eternal darkness, the ocean floor slowly spreads apart, tearing open the planet's crust.

Yet the scientists aboard the sturdy ship discovered astonishing vitality flourishing in that utterly hostile world more than two miles beneath the ocean's surface. *Alvin*'s lights reflected off slabs of iron-rich reds, bright yellows, and glittering flecks of pyrite scattered around rocky chimneys that belched 600-degree water colored black with minerals. Those hot springs, along with shimmering jets of compressed gases, supported a hierarchy of life. Bacteria, feeding on the materials rising from the vents, were grazed upon

SEEDS FROM SPACE

Gripped by temperatures near absolute zero, permeated by lethal radiation, airless and lonely—space could hardly seem more hostile to the delicate chemistry of life. And yet, in the view of many scientists, the seeds of life on Earth may in fact have been formed in those cold, interstellar reaches rather than in the warmth of Earth's primordial environment.

As early as the nineteenth century, respected scientists such as Germany's Hermann von Helmholtz proposed that organic molecules—the carbon-based compounds of which living things are made—were brought to Earth by meteoroids and comets. Although their presence is fleeting and fitful today (Comet West is seen crossing the skies in the photograph above), such chunks of rock and ice bombarded the planet early in its existence. The theory was heavily criticized ▶

Shooting 30,000 miles into space, jets of cyanogen gas envelop the rotating nucleus of Halley's comet in this false-color photograph, taken in April 1986 from an Australian observatory. Orange shows the highest concentration of cyanogen; yellow, green, blue, and black indicate decreasing levels of the gas.

Cyanogen, a simple organic molecule consisting of one carbon and one nitrogen atom, was an unexpected discovery in Halley. It may derive from the breakdown of more complex carbon molecules, called polymers, as the comet heats up on its approach to the Sun. By vaporizing some of the comet's ice, solar heat may also create the internal pressure that propels the swirling jets.

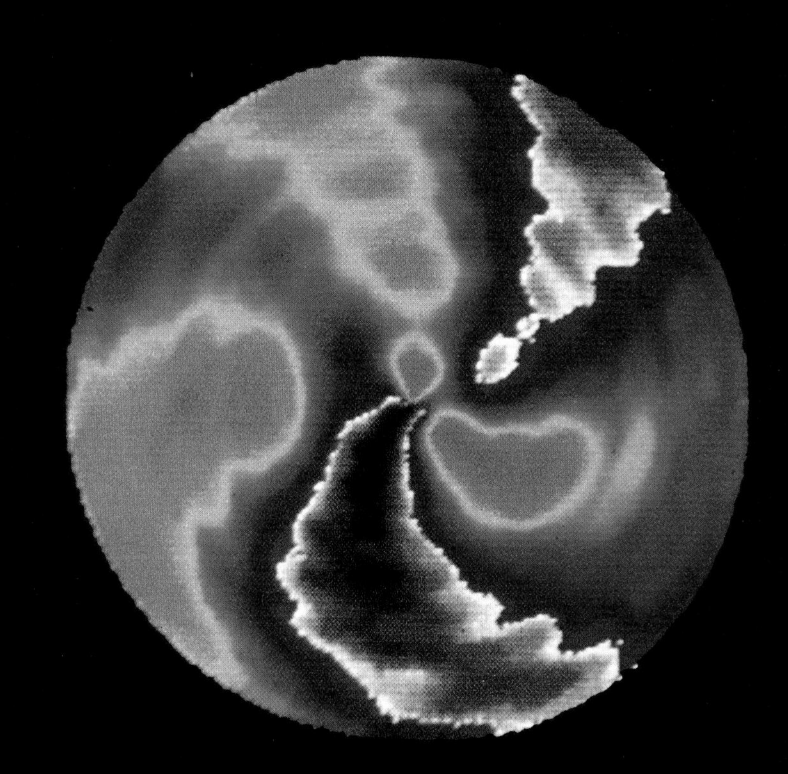

In a March 1986 probe of Halley's comet by the European Space Agency's Giotto spacecraft, instruments revealed the presence of formaldehyde polymers *(below)*, chemical chains of organic compounds linked by carbon (C) atoms. Although formaldehyde was the first polymer to be found on a comet, scientists believe it is just one of many such compounds on Halley. Computer simulations of Earth's early environment suggest that formaldehyde may have played a key role in the formation of the amino acids and carbohydrates needed for life.

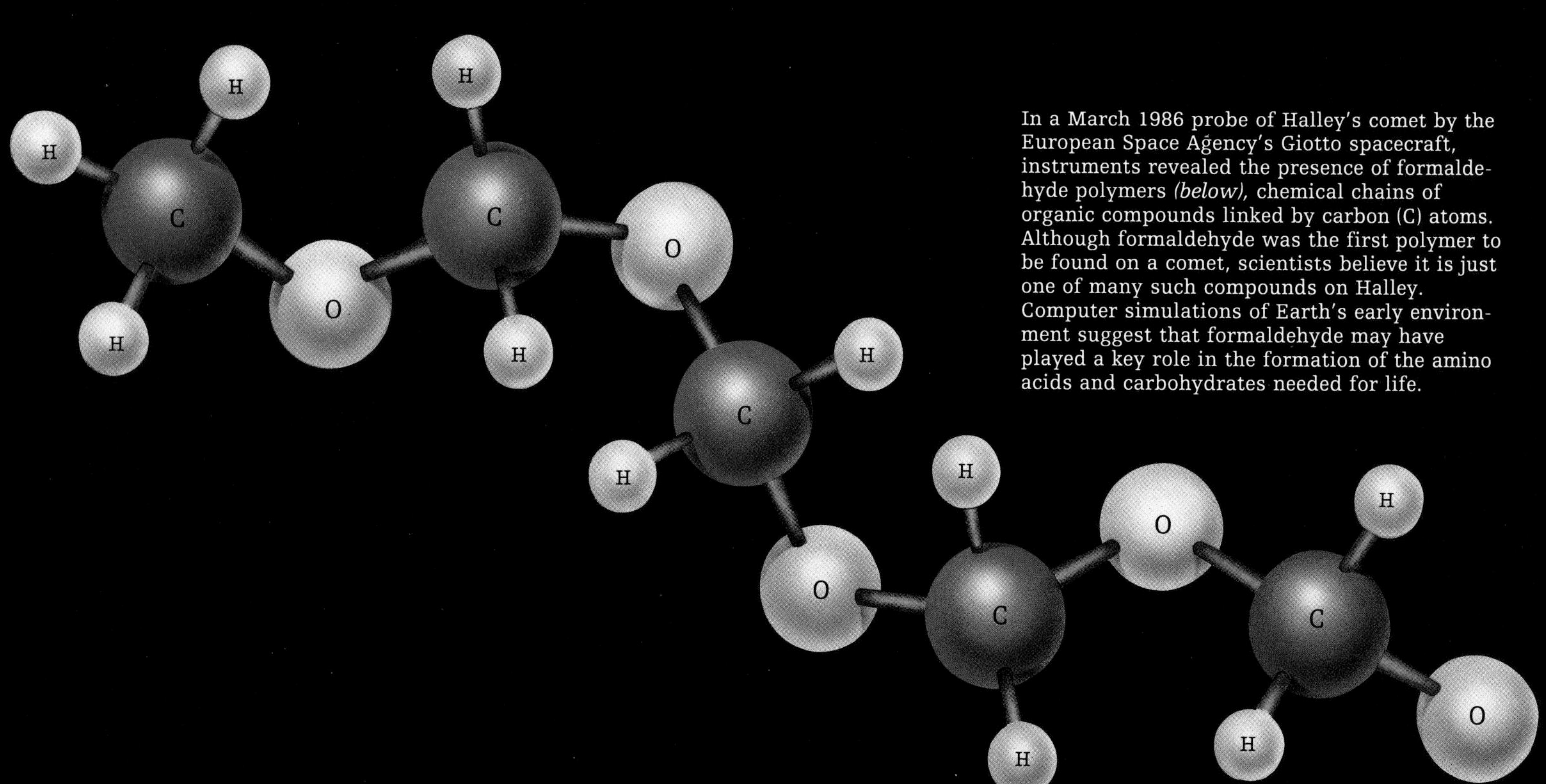

in its day, but it has gained a wider following in recent years. Instruments aboard spacecraft have studied Halley's comet *(top)*, and Earth-based telescopes have determined that it contains carbon-based substances. Similar molecules have been found in scrapings from meteorites *(opposite, top)*, as meteoroids that reach the surface are called.

Scientists are not sure how chemical evolution might work in space, but high-energy cosmic rays or heat from the birth of stars may cause collisions among atoms of hydrogen and helium within interstellar gas clouds, setting off a chain of events leading to the formation of complex molecules containing carbon, oxygen, and other, heavier elements. The heavier molecules may then clump together with dust to form the material of meteoroids and comets—wide-ranging travelers that may have scattered the elements of life throughout the Solar System.

atty acids extracted from a meteorite
hat fell near Murchison, Australia, in
969 produced a membranous substance
right) when mixed with chemicals that
nay have been present on primeval
Earth. Consisting of long hydrocarbon
chains, the molecules are the raw
naterials from which cell membranes
are built. Many scientists believe that
neteorites supplied the early Earth with
atty acids that could have turned into
ayered structures similar to cell walls.
According to one scenario, these acids
accumulated as foam on the oceans
and were carried by wind and waves
onto ancient tidal flats, where they
were formed into the cell membranes
of Earth's first microorganisms.

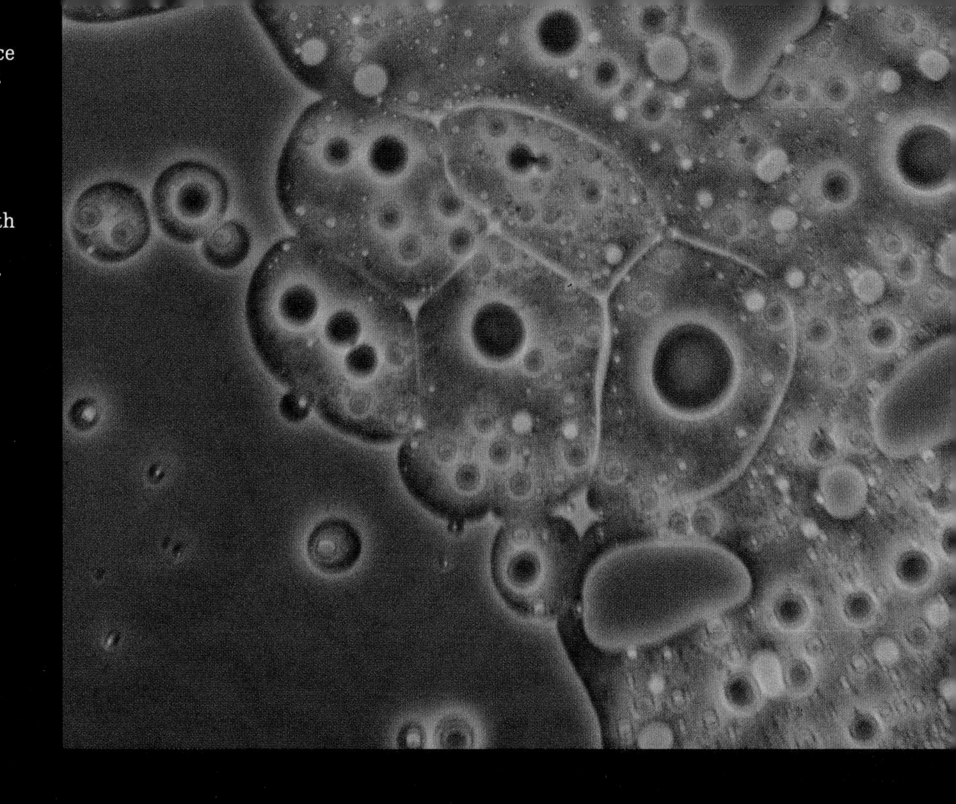

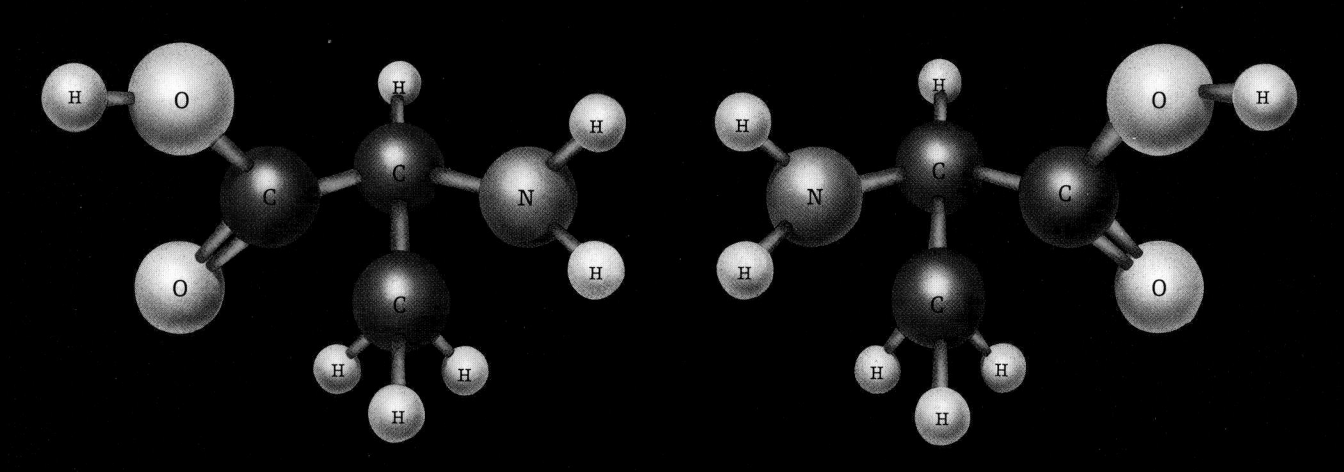

Among the organic substances found in the Murchison meteorite were molecules of the amino acid alanine *(above).* Made up of carbon (C), hydrogen (H), oxygen (O), and nitrogen (N), the molecules piqued scientific interest because they included about equal numbers of mirror-image "left-handed" *(left)* and "right-handed" *(right)* structures. When they are formed on Earth by biological processes, amino acids— the building blocks of protein—are almost always left-handed. The Murchison molecules therefore clearly could not be contaminants picked up on Earth; instead, they must have originated in space.

by blind shrimp and pale anemones, which in turn were food for white crabs.

Such a food chain has no parallel on the bright surface of the Earth. There, food chains are built on photosynthesis—the ability of many organisms to tap the power of solar radiation to convert nutrients into living tissue. But the underwater food web discovered by Rona and Edmond was built on the energy and sustenance that were drawn from the chemistry of the hot seeps along the rift. In this chemical-based ecosystem, the researchers found clams a full ten inches across, six-foot-long tube worms, and peculiar, disk-like oddities covered with a geometric pattern of dots resembling a tiny Chinese checkerboard. Rona learned later that the dotted creatures are identical to fossils found in 70-million-year-old marine sediments now exposed in the European Alps.

Half a decade earlier, another Alvin team, exploring similar submarine ridges near the Galápagos Islands in the Pacific, had been the first to find evidence of abundant life around "black smokers," as the dark-venting rock chimneys are called. John Corliss, a principal investigator on those dives, had been particularly fascinated by the teeming bacteria that metabolized hydrogen sulfide, ammonia, iron, and metallic manganese. For Corliss, a forty-one-year-old oceanographer then with Oregon State University, the resourceful bacteria represented the possible beginnings of life itself. Perhaps, he ventured, the rift communities are there because the rich mixing of elements, warmed and energized by the planet's internal fires, recreates the early stages of planetary evolution, a time when a world is first an incubator, and then a nursery. Although many scientists consider the superheated seafloor springs an unlikely and inhospitable cradle for the origin of life, there is general agreement that biology began somewhere, somehow, as chemistry.

EARTH'S GRAND EXPERIMENT

Life has been defined as that condition which distinguishes an organism from inert matter; telltale signs of its presence are growth, metabolism, reproduction, and adaptability. On Earth, matter began to stir with nonchemical vitality at some moment less than a billion years after the planet had coalesced out of the cloud of dust and gas that gave rise to the Solar System. Beginning with the simplest of creatures, which were not much more than animated molecules, life on Earth gradually moved toward ever-greater complexity, at ever-increasing rates.

Many times as earthly life progressed down a three-and-a-half-billion-year evolutionary trail, nature issued a cruel edict: Adapt or die. The first great crisis must have concerned energy. Early organisms obtained all the energy they needed from the chemicals of the primitive sea, but eventually the primordial soup began to thin. Some organisms then evolved an alternative to a chemical diet by drawing energy directly from the Sun. Not only did they save themselves, but their clever strategy also began to lay the framework for many life forms still to come. By extracting carbon from the carbon-dioxide-laden air of the young planet during the process of photosynthesis, these organisms inoculated the atmosphere with oxygen, a first, essential step in creating a terrestrial garden of life.

The fundamental vessel of evolution was also changing. Over millions of years, the rudimentary cell of living molecules, able only to divide and replicate, was transformed into a new kind of cell, containing the strands of genetic coding in a nucleus. Its arrival signaled the beginning of sexual self-replication, passing hereditary messages of increasing sophistication from one generation to the next. About a billion years ago, the first multicelled organisms appeared: Seaweed, segmented worms, medusas, and fish spread through the sea.

LIFE'S LONG JOURNEY

Roughly 10 billion years after the genesis of the universe, the Earth was born out of a cloud of dust and gas drifting in the outer reaches of the Milky Way galaxy. Another 4.6 billion years would bring the history of the Solar System's third planet forward to the present day. Much of that history has been bound up with biological evolution, a long and complicated process whose major developments are illustrated below and on the following pages. The evolutionary record is critical to the search for extraterrestrial life: Earth's past holds our only clues to how, why, or when living creatures might arise elsewhere.

In one common scenario, Earth life began about four billion years ago, when volcanism *(below)* spewed forth gases to create a thick atmosphere and water vapor clouds condensed to form the oceans. Crashing meteorites and comets endowed the young Earth with organic molecules that would later gather in living assemblages. Electrical storms, volcanism, and radioactivity provided ample energy for life's inception.

Single-celled microorganisms emerged in Earth's seas by 3.5 billion years ago. Subject to random, often tiny changes called mutations, and aided by nature's tendency to favor helpful mutations, the one-cells eventually evolved into rudimentary multicelled organisms. Forms such as vertebrate animals appeared just 500 million years ago, about 100 million years before plants and animals first took to the land.

As oceans rose and fell and continents drifted, new environmental conditions fostered new modes of life, leading to the extinction of previously dominant species. One such crisis—its exact nature unknown—felled the dinosaurs. Their demise was followed by the rise 65 million years ago of a newly dominant class, the mammals, and eventually by intelligent human life in its present-day form. If the environment had evolved differently, or if the dinosaurs had managed to adapt and survive, other life forms may well have emerged. On such an Earth, the authors and readers of this book might not even be mammalian.

4 billion years ago

3.9 billion

3.8 billion

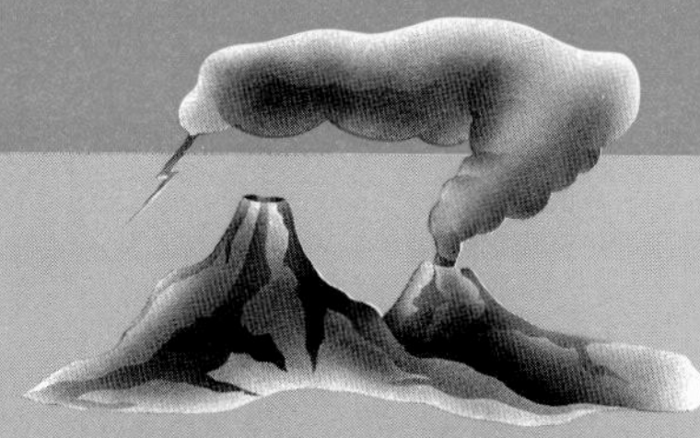

HADEAN EON

ARCHEAN EON

Setting the stage. Half a billion years after its birth, the infant Earth writhes in volcanic eruptions. Expelled gases contribute to a thick atmosphere of carbon dioxide, nitrogen, and carbon monoxide, with traces of ammonia, methane, and hydrogen sulfide.

Then the land areas of the continents, which had lain almost barren under a barrage of unfiltered ultraviolet radiation from the Sun, became gradually habitable. Some twenty miles above the planet's surface, an interaction between sunlight and oxygen in the atmosphere began to forge an invisible barrier. Ripped apart by the high-energy action of short-wavelength ultraviolet rays, the normal two-atom variety of free oxygen recombined in a highly reactive triatomic form called ozone, which absorbs the harsher forms of ultraviolet radiation. Some half billion years ago, as the lethal bombardment was lifted, a few enterprising creatures moved tentatively from the ocean to the land.

A WORLD OF GIANTS

From this figurative instant in geologic time, the great exfoliation of life on Earth began. Species radiated outward in all directions, to fill every nook and cranny, driven by the unyielding impetus of natural selection to replicate, interact, and alter through genetic mutation. The great lizards called dinosaurs commenced their 140-million-year reign, only to vanish, somehow destroyed in the coils of the very evolutionary process that spawned them. By trial and error, epoch by epoch, the Earth was populated by resilient species. The skies filled with the improbable aerodynamic shapes of bees and butterflies, the marginally flightworthy form of a flying dinosaur, and finally with multitudes of warm-blooded birds.

Every creature had adapted to its world in some fantastic way. Aquatic eels and rays evolved the ability to stun their prey with electric shocks; spiders spun webs they could cast like gladiators' nets; the voracious sperm whale hunted the deep abysses of the sea. Earth became a world of giants, like the 300-foot-high sequoias whose seeds first germinated more than 60 million years ago, or 135-ton blue whales whose ancestors 70 million years ago forsook the land for the plenum of the sea. But even with its titans, Earth—past and present—may be owned by the very small. Microbes, the tiny organisms consisting of only a cell or two, seem capable of adapting to almost any condition or circumstance. Bacteria flourish everywhere. Plankton, the simple plants at the base of the oceanic food web, appear to represent a kind of fundamental life form, one that helped transform the primitive atmosphere of early Earth. Viruses, which seem to straddle the line between life and

3.7 billion years ago

3.6 billion

nonlife, comprise only a few molecules and yet, through their coopting of cells, in some ways rule the world.

Today the planet supports a total biomass of several trillion tons, distributed among an estimated 300,000 species of plants and a million species of animals. This is life on Earth—stupendous testimony to the power of matter to order itself and find opportunities at every turn. But its beginnings remain a mystery, and so, therefore, do the probabilities of life's launching itself into existence elsewhere. At the root of the matter, Earth life and human dreams of companionship in the universe raise the same fundamental question: What is the spark that transforms the reactions of mere chemistry into living biology? Until the nineteenth century, scientists and laypersons alike would have thrown up their hands: Life seemed, somehow, to come from nowhere.

SOMETHING IN THE AIR

Take one tub of grain, add one dirty shirt, and let stand for three weeks. This seventeenth-century recipe for mice was formulated by the Belgian chemist Jan Baptista van Helmont, who, twenty-one days after adding shirt to grain, was rewarded with a tubful of live mice. His prescription was nothing new. For more than twenty centuries, naturalists had believed in a kind of spontaneous generation of life: Fleas and lice came from the slime of rivers and wells, crabs from seaside decay, and many kinds of fish from rotting seaweeds. But doubts crept in as microscopes became more and more powerful. Life seemed to permeate the world: A drop of water from the Thames or the Seine was observed to be more populous than the great cities splayed on their banks.

The theory of spontaneous generation was laid to rest in the nineteenth century through the efforts of French chemist Louis Pasteur and Irish physicist John Tyndall. In a series of experiments, Pasteur demonstrated that boiled broth sealed in a glass flask remained a lifeless pale brown liquid. If opened to the atmosphere, however, it soon teemed with microorganisms. Tyndall showed why. His interest in the light-scattering properties of dust led him to conclude that some airborne particles were microorganisms. But his work took him beyond the idea of minuscule life forms drifting in the air, to a conceptual leap that linked microbial life

The emergence of simple life forms. The earliest known living beings were water-dwelling one-celled microorganisms, like the types shown below at left. The genetic molecule DNA enabled each to replicate through simple cell division. Feeding on carbon-rich compounds, the organisms spread in layers to form silt-filled mats. The mats in turn created fossilized forms called stromatolites *(below, right).*

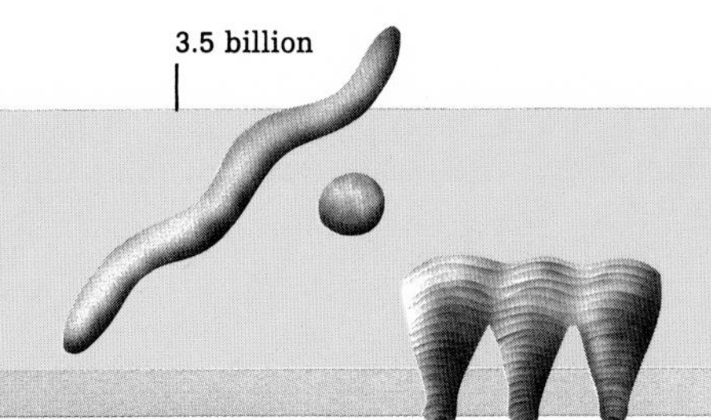

to the inanimate, physical world of chemistry. "I cross the boundary of the experimental evidence," he told the British Association for the Advancement of Science in Belfast in 1874, "and discern in that matter . . . the promise and potency of all terrestrial life."

Gradually other scientists fastened upon the idea that biology sprang from chemistry. In 1922 Aleksandr Oparin, a twenty-eight-year-old Soviet biochemist, experimented with mixing small proteins called histones and an emulsion of gum arabic into water. He found that they formed a stable droplet chemists call a coacervate, and he noted that the coacervate loosely resembled the framework of a cell. Oparin proposed that the Sun's energy, in the form of ultraviolet radiation, could have spurred such a self-assembly of life-building compounds in the young planet's seas.

At about the same time, British geneticist John Haldane was also drawn to the notion that solar radiation powered the leap from chemical to organic life. He had been fascinated by life sciences since the age of eight, when he first assisted the researches of his father, a noted Oxford physiologist. Haldane shared Oparin's political as well as scientific convictions, and was for years an editor of the Communist party's *Daily Worker.* In a paper published in 1929, he imagined that terrestrial life began on a primitive, hostile Earth whose atmosphere contained no free oxygen—no oxygen, that is, in its two-atom molecular form. Incoming ultraviolet radiation would have caused the formation of prebiological molecules in a primitive organic soup, and they would subsequently have clustered into primitive cells. These first tiny organisms were anaerobic; that is, they did not need oxygen to metabolize nutrients. (Anaerobic organisms still exist: Bacteria can grow in sealed cans of food.) "We should expect," Haldane wrote in 1929, "that high organisms like ourselves would start life as anaerobic beings, just as we start as single cells."

For many years, the general community of biologists found Oparin's and Haldane's complementary hypotheses too improbable to take seriously. Then, in 1953, two American chemists put the ideas to the test.

THE PRIMORDIAL FLASK

Harold Clayton Urey, a well-established chemist, was a relative newcomer to the study of life's origins on Earth. He had won the 1934 Nobel prize

3.2 billion years ago

3.1 billion

in chemistry for his discovery of deuterium, the heavy form of hydrogen later involved in nuclear weapons research. But Urey, the son of an Indiana lay minister, was a man of strong personal ethics. In 1945, dismayed that his achievement had been used for nuclear weapons development, he joined the faculty of the University of Chicago, determined to do no scientific work that could be connected to military technology. The field he turned to was Earth sciences. There, the scientist focused on reconstructing a temperature history of the primeval oceans. Peering into the planet's distant past led him to an increasing interest in the beginnings of terrestrial life.

In 1953 Stanley Miller, one of Urey's graduate students, found himself without a proper dissertation topic for his Ph.D. With encouragement from Urey, the twenty-three-year-old Californian decided to build a laboratory version of the primitive Earth—a prototype of any young planet with a warm sea and an oxygen-free atmosphere.

Miller re-created this world as a set of glass tubes and spheres. At the bottom of the apparatus, he filled one sphere with water and heated it to the boiling point. The rising steam, representing a speeded-up version of evaporation from the ancient oceans, climbed through glass tubing to a second sphere, containing methane, ammonia, and hydrogen in its two-atom, or molecular, form—a simulation of the early atmosphere. As the steam swirled into the upper flask like rising clouds, Miller generated 60,000-volt sparks to zap the mixture with the laboratory equivalent of lightning.

The steam-borne gases then passed through a cooling chamber, where the water condensed into droplets that ran down the tubing and collected in a glass U-bend. The two scientists settled in to wait for their primitive world to make the next move. After about twenty-four hours, the condensed water turned pink and then darkened to red. Chemical analysis told exactly what the

The oxygen revolution. More and more bacteria photosynthesize, combining solar energy with carbon dioxide and hydrogen to produce energy-rich sugars. Anaerobic bacteria —those not adapted to oxygen—extract the hydrogen from atmospheric hydrogen sulfide. A minority of aerobic, oxygen-tolerant bacteria acquire hydrogen by breaking apart water molecules, releasing oxygen as a by-product. The freed oxygen creates iron oxide sediments like the rusty bands in the rock below.

3 billion

2.9 billion

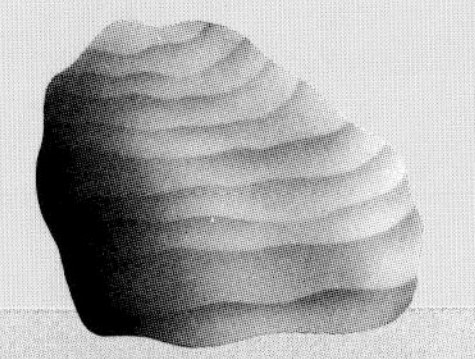

deepening color meant: Organic molecules had been formed by the reaction of electrical sparks and gases. In the microcosm of collected water, Miller and Urey found several simple amino acids, the building blocks of proteins, as well as indications of more complex compounds. The homemade primordial brew also contained sugars, hydrogen cyanide, and formaldehyde. In all, about 15 percent of the carbon that makes up methane had been converted into organic compounds.

At first, their results seemed to confirm the view that life was an accident of random chemical interactions. But, despite a rich assortment of the rudimentary chemicals of biology, their soup remained inert, failing to draw energy from its surroundings or replicate its makings in any way. Although the Miller-Urey experiment caused a considerable stir in the scientific community, it failed to identify the impulse that animates the inanimate.

BIOLOGY'S MESSENGER

While Miller and Urey were tackling the problem through chemistry, biologists were coming at it from the opposite side of the biochemical frontier. In looking at the way life forms reproduce, they zeroed in on genes, little-understood cellular components that were surmised to steer the complex process of heredity. By midcentury, biochemists in Europe and America had determined that genes were composed of the chemical substance deoxyribonucleic acid. But the structure and real function of DNA were not known until 1953, when British biochemist Francis Crick and an American colleague, James Watson, cracked the code.

Working together at England's Cambridge University, the young researchers determined that DNA took the form of a two-stranded coil—what came to be known as the double helix—on which living things stored their genetic information. The double-spiral framework of genes explained how life ensures its continuity, passing its instructions on from generation to generation *(pages 32-37)*. As one scientist put it, the discovery marked "the appearance of biological organization out of molecular chaos." Once the surprising predictability and order of DNA's genetic role became clear, investigators began to think this coding apparatus must have formed very early in life's history. Perhaps, some believed, the first appearance of the DNA molecule marked the moment when chemicals became life.

2.8 billion years ago

2.7 billion

2.6 billion

ARCHEAN EON (cont.)

To others, however, the double helix seemed too complex to represent the start of evolution; DNA looked more like an end product. In 1958 Sidney Walter Fox, a California-born biochemist who was then working at Florida State University, came to believe that the answer to the question of life's origins lay in proteins or, more particularly, in the way these compounds assemble themselves into cell-like structures. Proteins, according to Fox, must have come before DNA.

In a sense, Fox could be considered a modern counterpart to Aleksandr Oparin. Like the Soviet scientist, Fox thought that a prototypical cellular structure could have helped concentrate the right arrangement of chemicals. In Fox's view, however, an intermediate, protein-forming step was required. His own experiments showed that amino acids, of a type that might have trickled out of a primitive atmosphere, responded remarkably well to heat, ordering themselves into proteinlike molecules. Put into water, these proteinoid molecules grouped into tiny spheres with a notable resemblance to living cells. They divided when new protein was added to the water, and they even seemed to mimic the electrochemical dialogue that takes place in nerve cells. Fox said these "microspheres" may in fact have been primitive cells, and he speculated that such entities could have formed on a young Earth of steamy oceans and active volcanoes.

Other biologists were dissatisfied with both views: With its genetic apparatus, DNA seemed too elaborate a starting point, but proteins, even in conveniently walled packages, seemed to carry too little hereditary information to constitute life's first biochemical step. The Sinhalese-American chemist Cyril Andrew Ponnamperuma observed that a structure unable to re-create itself was not really alive in any case. Any primitive form of life, he believed, would have required some kind of genetic system in order to replicate; otherwise it was only interesting chemistry. In the early 1960s, Ponnamperuma, then at NASA's Ames Research Center, began devoting himself to finding this prototype genetic system.

Ponnamperuma hypothesized that life might have begun as a molecule of

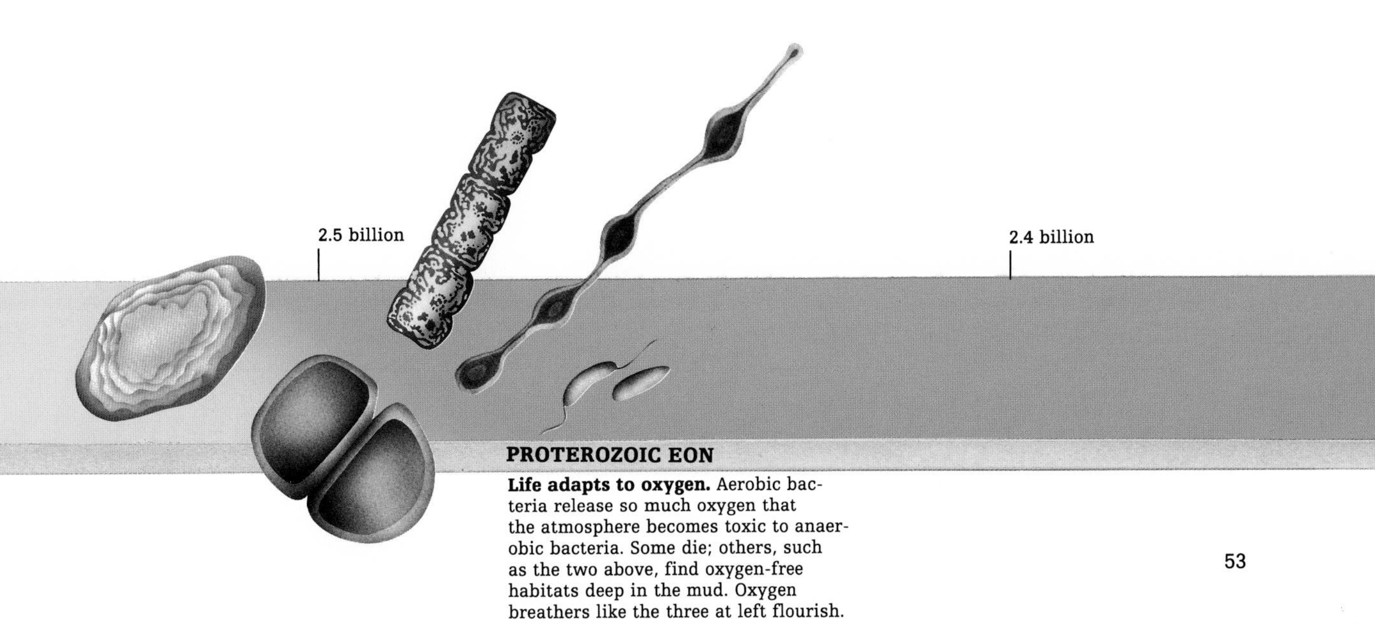

PROTEROZOIC EON

Life adapts to oxygen. Aerobic bacteria release so much oxygen that the atmosphere becomes toxic to anaerobic bacteria. Some die; others, such as the two above, find oxygen-free habitats deep in the mud. Oxygen breathers like the three at left flourish.

ribonucleic acid, or RNA, a single-stranded nucleic acid that acts as a messenger between DNA and the cellular material that assembles proteins. In the earliest stages of biological evolution, RNA might have been able to make crude copies of itself and to interact readily in a world pervaded by strong chemical activity. In more sophisticated variations on the earlier Miller-Urey tests, he exposed various atmospheres to bombardment by a powerful stream of electrons, simulating high-energy cosmic rays. When Ponnamperuma analyzed the atmospheres afterward, he found adenine, one of the handful of chemical constituents of nucleic acids that carry the coded instructions in DNA and RNA. The chemist concluded that intense radiation could have squeezed these compounds, called bases, out of the atmosphere of the young Earth.

The next step—the biological organization of the ingredients—was simulated by a European team of researchers in Göttingen, halfway between Bonn and Berlin. During the 1970s and 1980s, Manfred Eigen and his colleagues at the Max Planck Institute for Physical Chemistry demonstrated that nucleotides, the chemical building blocks of RNA, could be transformed into RNA in a kind of copying process: In the presence of an existing molecule of RNA that served as a model, a protein catalyzed, or promoted, replication. Further research revealed that nucleotides could be catalyzed into RNA even when they had no RNA molecule to copy. This RNA formed less predictably and more slowly than molecules that could copy a model, but once the process had begun, it proceeded as usual. Moreover, the researchers found that some of the RNA produced in this way would undergo a genetic change—a mutation—that made it reproduce itself more efficiently. The experiments just skirted the leap from chemistry to life.

Meanwhile, at the Salk Institute, located in La Jolla, California, Oxford-trained Leslie Orgel and his coworkers were turning Eigen's experiment inside out: Their findings showed that a strand of RNA can, under the right chemical circumstances, make a copy of itself without a catalyzing protein, or enzyme, being present. Orgel used the element zinc as a catalyst to encourage the formation of long chains of RNA, without any help from the enzymes that keep living things functioning today.

By the late 1980s, the issue was not whether organic chemicals could have assembled into life but, rather, how they were drawn or driven together.

Researchers put forward a variety of possible engines to power the crucial reactions: ultraviolet radiation, lightning, heat seeping from the Earth's center, the crash of meteorites, even the beat of ocean waves on rock and sand. Some scientists believed such cookery took place in the atmosphere, which rained premixed organic compounds on the surface. Others offered a more down-to-earth solution.

MOTHER CLAY

University of Glasgow chemist A. Graham Cairns-Smith found RNA, like DNA, "too high tech" to have been a starting point for life. "The first organisms," he wrote in 1985, "must have been 'low tech' machines of the kind that are fairly easily put together and for which there are simple versions that work, more or less (spears, not machine guns)." Cairns-Smith suggested that the present essential elements of biology—nucleic acids and catalyzing proteins—evolved on a kind of prebiotic scaffolding. Ultimately the scaffolding disappeared, leaving behind the biochemistry that now controls the processes of all living things. The vanished supporting structure, in the view of Cairns-Smith, could have been clay.

In his imagined prebiotic world, tiny pores formed between the soil particles that, over the ages, were compressed into sedimentary rock. These pores, in material like sandstone, then became conduits for solutions of water and dissolved minerals produced by eons of weathering. Within the pores, thin layers of an odd, malleable crystalline material—clay—began to form, "learning" to hold themselves in place by developing certain structural defects. For example, the young clays may have evolved a light, permeable structure to allow water to pass through without washing the frail layers away. According to Cairns-Smith, the clays developed a rudimentary ability to reproduce these adaptations as the mineral crystals grew, copying not only the original structure but the defects as well. Although there was no life, there was the evolution of what he called "mineral crystal genes." As the clays

A primitive partnership. Large soft-membraned bacteria engulf smaller ones *(below)*, which then process incoming oxygen fuel for their host. No longer considered cells, the small respiration managers evolve into mitochondria, one of many such biochemical helpers, or organelles, found in both plant and animal cells.

2.1 billion

2 billion

1.9 billion

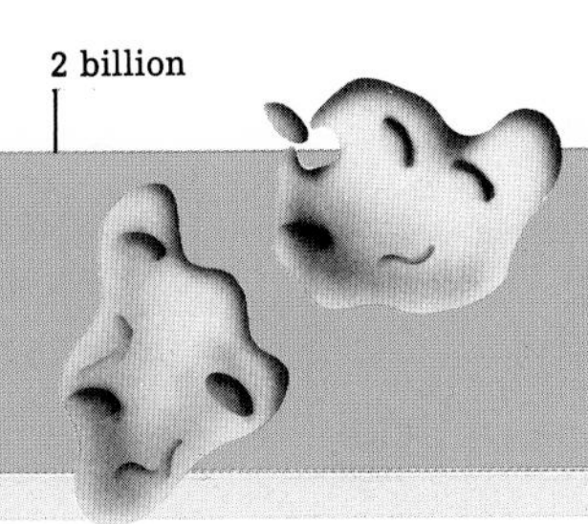

continued to evolve over millions of years, the organic precursors of RNA and proteins appeared, at first playing only a very minor role in the business of replication and the conversion of energy to growth. But, supported by their clay scaffolding, the newcomers were able to evolve into extremely complex organic forms, which gradually took over the copying and message-sending work formerly done by clay. Over still more eons of evolution, they completely replaced the geochemical genetic material.

Experiments offer some support to Cairns-Smith's highly speculative theory. X-ray studies of clays show not only that their crystals may change shape to grow into a new environment but also that they seem to pass the new order on to succeeding crystals. More intriguing is evidence that organic molecules tend to cluster readily on the crystalline surfaces of clays. Researchers have shown that wet-dry cycles in clay actually promote the assembly of amino acids. One scientist, Lelia Coyne of California's San Jose State University, has taken the idea of damp and drought even further. In a series of experiments, Coyne soaked clays in liquids containing organic solvents and exposed them to various types of radiation. When these clays were subsequently shocked, fractured, or resoaked, they emitted the kind of radiation to which they had been exposed. "If you hit it with a hammer it blows ultraviolet energy for a month," she explained, adding that the ability to store and release energy, like the transmission of genetic information, is a basic property of life.

If Cairns-Smith and other clay theorists are correct, carbon-based life on Earth spent its infancy in a kind of clay cradle that could be common elsewhere in the Solar System. For example, clays are found, along with amino acids and other organic compounds, inside meteorites *(pages 43-45)*, and the results of Viking's soil analyses suggest that the surface of Mars is dominated by layered clays rich in iron. Because such minerals are produced on Earth by the weathering action of a dense atmosphere and long exposure to liquid water, some scientists believe the Martian clays may be the mineral remains of wetter times on the Red Planet.

ORGANISM EARTH

The clay cradle envisioned by Cairns-Smith may be only a detail in the canvas of a broader and more controversial hypothesis known as the Gaia theory,

1.8 billion years ago

1.7 billion

after the Earth goddess of Greek mythology. According to the Gaia view, life shapes a world—this or any other—to suit itself. The theory was developed in the 1970s by British scientist James Lovelock. Although his degree is in medicine, he has made his living mostly as an inventor of scientific instruments, working independently in a rustic laboratory at Coombe Mill, overlooking the valley of the river Tamar as it crosses Devonshire toward the sea. He has taught at many universities, including Yale and Harvard in the United States. But it was as a member of Norman Horowitz's biosciences team at the Jet Propulsion Laboratory in the late 1960s that Lovelock first saw the rough shape of the theory he would later introduce with the provocative phrase "the Earth is alive."

In the United States, Lovelock gained a determined supporter in Lynn Margulis, a Boston microbiologist who had independently come to the idea of a self-regulating Earth. The two coauthored many of the arguments for Gaia, gradually marshaling physical evidence for the intricate interconnectedness of life and environment. One of the studies they cited, for example, showed that marine phytoplankton release a sulfur compound that migrates into the atmosphere as particles. Water quickly condenses around the particles, encouraging the formation of clouds. Increased cloud cover in turn shades the ocean, cooling its surface layers to the temperature preferred by the plankton.

There are many other examples of such Gaian interaction. Perhaps the most compelling is the way in which Earth's atmosphere was transformed some two billion years ago from one almost completely devoid of oxygen to the present, richly oxygenated mantle of air. Scientists generally agree that the appearance of photosynthesizing bacteria and algae, drawing in carbon dioxide and emitting oxygen, marked the beginning of the atmosphere that sustains most of the planet's life today. In other words, early life created the conditions required for the further development of life. But the example also

Sex is invented. Cells that are called eucaryotes *(below, right)* evolve a membrane-enclosed nucleus, where genetic material is stored, allowing eucaryotes to replace self-replication with the union of two cells. Since each contributes material to yield a third unique cell, genetic variety in living things increases dramatically. Nonnucleus cells, or procaryotes *(below, left)*, persist but evolve more slowly.

Greening of the Earth. Still confined to the water, bacterial life has by now single-handedly transformed Earth from a barren volcanic environment to a fertile garden. In some eucaryotes, a new type of organelle called the chloroplast appears, specializing in capturing sunlight through the process of photosynthesis. By definition, cells that are equipped with chloroplasts are plant cells.

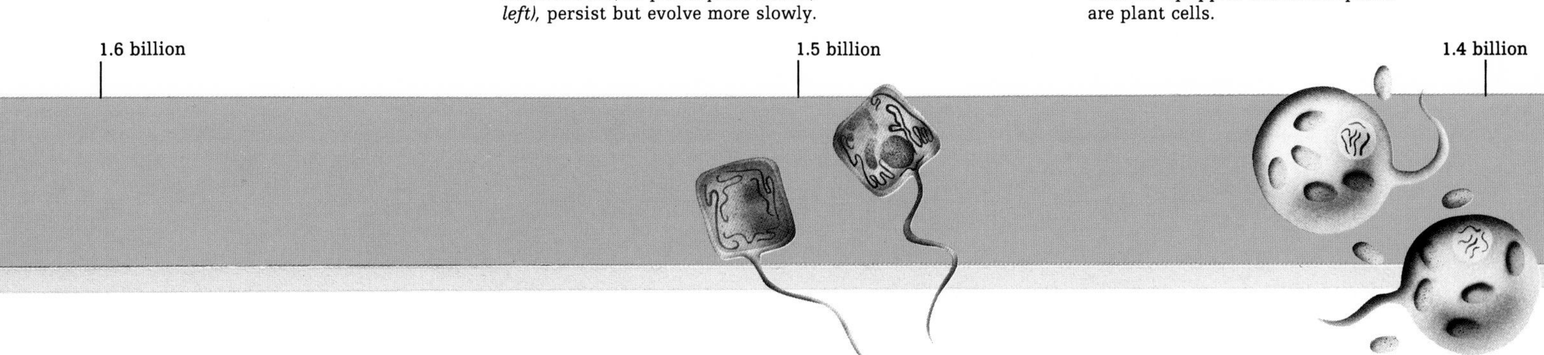

reveals that there is nothing sentimental about Gaia: Oxygen is a deadly poison to organisms that have not adapted to its use. "All organisms on the face of the earth were threatened by this gaseous oxygen," Margulis noted in 1984. "Some bacteria simply crawled into the muds, but others evolved mechanisms for dealing with oxygen."

Gaians also note subtler intertwinings of life and biosphere. For example, although the Sun's luminosity has increased significantly during the Earth's lifetime, surface temperatures have not. In Margulis's view, since different microbes produce different gases, varieties of microorganisms evolved to produce methane and other hydrocarbon compounds that helped modulate terrestrial temperatures.

DAISIES IN A MODEL WORLD

A mathematical simulation, or model, developed by Lovelock approximates the self-regulating cycles of Gaia. Daisyworld, as he called the model, is a planet whose environment has been narrowed to the single parameter of temperature, which varies with how well the world reflects or absorbs solar radiation. If the planet is relatively dark, it absorbs more solar radiation and its temperature rises. If relatively light, it reflects more sunlight and cools itself. As the name implies, Daisyworld is populated solely by daisies, some of them dark, some light.

Set into imaginary motion billions of years in the past, the first generation of daisies would have been equally distributed among dark and light flowers. But in those early epochs, the Sun was slightly cooler than it is today, and the planet relatively chilly. The dark daisies would have absorbed more sunlight and warmed themselves, thriving on their cool planet. The light daisies, on the other hand, cooled themselves by reflecting sunlight; on a cold world, this would have worked against them, causing them to thin. Gradually, dark daisies would have become the dominant life form on Daisyworld. But then, as the Sun brightened with time, the dark daisies would have found the

The eucaryotes diversify. One-cell eucaryotes called protists (for "first") develop into a variety of more complicated forms. Tiny plants and animals, the protists include amebas and some forms of algae and plankton. Many acquire lashing tails or wriggling hairs, enhancing their mobility.

1.3 billion years ago

1.2 billion

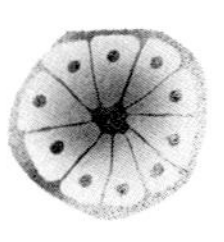

radiation-absorbing property that had caused them to flourish now working against them, creating a world too hot for comfort. However, this warmer world would be perfect for the self-cooling white daisies, which would slowly displace their dark cousins across the vast, mathematically created fields of Lovelock's world. Again, their growing abundance would increase the planet's reflectivity and cool it, re-creating the world in which dark daisies could succeed. The cycles of Daisyworld are meant to represent, in greatly simplified form, those of Gaia, which is, in ways that are still mystifyingly complex, similarly self-regulating.

The Gaia theory does not assume that Earth is predisposed in favor of one form of life over another. Just as the advent of an oxygenated atmosphere doomed most of the anaerobic life forms then on the planet, the industrial advance of humankind, which loads the atmosphere with heat-trapping gases and razes oxygen-giving forests, may be destroying the very machinery that sustains life on Earth. Margulis has emphasized that the organism that is Earth might not necessarily take corrective action to protect human life, especially if the civilization humans have created threatens the whole biosphere. In her view, survival would go to the microbe, which she believes is the living foundation of all other life. "Kill off animals and plants and the planet will recover," she says, "but kill off microbes and in weeks the earth will be just as sterile as the moon."

The apparent sterility of space squares with Lovelock's expectations. While at JPL, he argued that certain atmospheric characteristics indicated life, at least as we know it, and others did not. An oxygen-rich atmosphere—especially one with a radiation-absorbing shield of ozone at high altitudes *(page 132)*—would be a clear life signal, he believes. In contrast, atmospheres choked with carbon dioxide and stagnating for lack of the vertical mixing that comes with weather systems probably shroud barren globes. "Long before Viking set course from Earth," Lovelock wrote a dozen years after the 1976 mission, "I felt intuitively that life could not exist on a planet sparsely; it could not hang on in a few oases, except at the beginning or at the end of its tenure. As Gaia theory developed, this intuition grew; now I view it as a fact." To the British inventor, either planets are alive or they are not. As on Daisyworld, life in abundance is required if there is to be life at all.

Thus far, the universe has yielded not a single sign of extraterrestrial

1.1 billion | 1 billion | 900 million

life, either in abundance on some Gaia-like planet or in tiny pockets of sheltered vitality. And yet, scientists have found the universe richly endowed with chemicals from which terrestrial life must have risen. If life began on Earth as a product of random chemistry, it should have begun elsewhere in the same way. That it does not appear to have done so—at least in our immediate vicinity—leads some scientists to suggest that the first chemical steps toward life on Earth were, first, not random at all, and second, not of earthly origin. Perhaps the seeds of life were scattered by some unknown, extraterrestrial hand.

SEEDS EVERYWHERE

That possibility was vigorously explored early in this century by the Swedish chemist Svante Arrhenius, a widely respected scientist who had won a Nobel for his studies of chemical reactions. In 1908, at the age of forty-nine, Arrhenius published *Worlds in the Making,* devoted entirely to the notion that Earth had lain fallow until planted with seeds from outer space. He proposed that simple living forms, spores, escaped from the atmospheres of life-bearing planets and drifted from world to world, transmitting the spark of life. Arrhenius called it panspermia—seeds everywhere.

The panspermia thesis has attracted more than its share of top minds, both before and after Arrhenius coined the term. For example, the British physicist William Thomson, Lord Kelvin, gingerly broached the idea in 1871, in an address to the British Association for the Advancement of Science. "The hypothesis that life originated on Earth through moss-grown fragments from the ruins of another world may seem wild and visionary," he said. "All I maintain is that it is not unscientific." A century later, British astronomer Fred Hoyle, known for his studies of the nuclear interior of stars, threw his considerable reputation behind the concept of life-bearing spores adrift in space. Working with Chandra Wickramasinghe, a colleague from India, Hoyle promoted the view that perhaps 80 percent of interstellar dust is composed of planet-hopping bacterial and algal cells.

Biologists scoffed at the possibility, pointing out that the small seed organisms would be exposed to lethal doses of radiation during space travel. Further, they argued, if such spores had been escaping from the Earth, some should have settled on the Moon, which is known to be barren. An alternative,

Cradled by the sea. With life still keeping to its comfortable water habitat, the first known multicellular animals appear. They range from jellyfish and segmented worms to coral-like sea pens. The newcomers give rise to a great multiplicity of species—some, like the jellyfish, still thriving in present-day seas.

800 million years ago

700 million

dubbed "directed panspermia," was offered by DNA code breaker Francis Crick and California chemist Leslie Orgel. In 1973 they proposed that a distant civilization might deliberately sow the galaxy with the seeds of life, sending out the life-carrying materials encapsulated in spaceships to protect them from deadly radiation.

The notion of organisms traversing the void of space, whatever the mode of disbursement, raises a question central to the search for life in other parts of the universe: What sort of conditions can living matter endure on foreign worlds? The nearly undetectable creatures of Antarctica, the sunless ecosystems of the ocean floor, the tiny animals whose lives are suspended between infrequent sips of moisture—all have sent a message to terrestrial biologists: Rules that imply limits to the resilience of life are made to be broken. Perhaps, some scientists believe, it is time to redefine life itself to fit these broader possibilities.

ENERGY AND ORDER

In the view of Gerald Feinberg, a theoretical physicist at Columbia University, and Robert Shapiro, a chemistry professor and geneticist at New York University, terrestrial exceptions to conventional definitions of life abound. For example, the microscopic animals called tardigrades, when dormant, are as inert as ash and do not reproduce, metabolize, or grow. But, given a revitalizing splash of water, they bounce back to life. Such sterile hybrids as mules cannot reproduce, but are alive and kicking. In *Life beyond Earth,* published in 1980, Feinberg and Shapiro urged instead a redefinition that could be applied to all forms of life, everywhere. While their ideas braided together such disparate theories as Gaia and panspermia and eased some contradictions among physics, chemistry, and biology, the Feinberg-Shapiro perspective was a kind of exobiological shock wave disturbing the tranquil medium of biology as usual.

Life, they proposed, springs from interactions occurring between free energy and matter capable of using it to increase order—nonrandomness—within the interacting system. The energy might go toward building cell walls. It might be employed to organize the matter inside an egg into an embryonic chick. The resulting expressions of life, in turn, would use more available energy to create still more orderliness, and so on, as long as con-

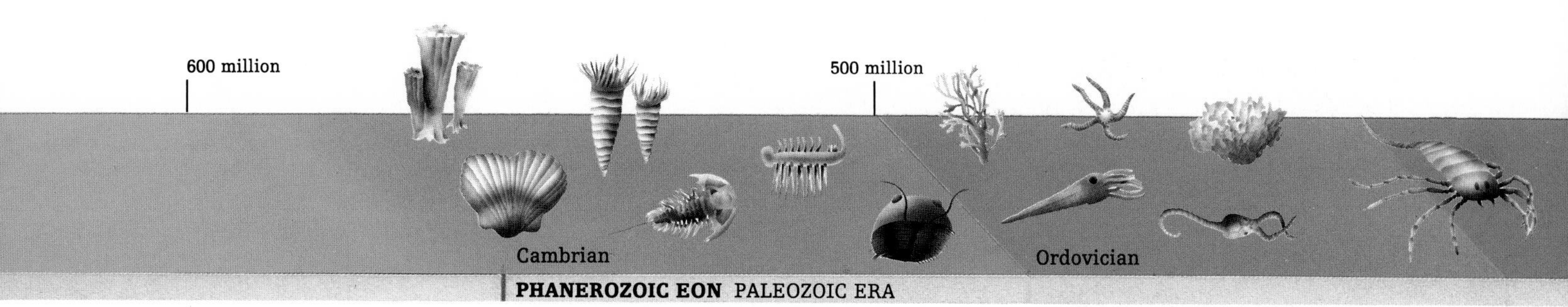

Adding outer protection. Marine life grows more various still, producing the first animals with hard outer coverings *(above).*

Introducing the vertebra. With the advent of primitive fish, the animal kingdom splits into the vertebrates (animals with backbones) and invertebrates (those without). Invertebrates like the nine-foot-long sea scorpions *(above, right)* dominate the oceans.

ditions permitted. Feinberg and Shapiro likened the flow of energy through matter to the movement of money in a poker game, which can last only as long as there is money still to be bet among the players. When there is not, the game has reached the point of equilibrium, a stagnant state in which no further order—no life—is possible.

On Earth, the main flow of energy is solar radiation, which passes through and energizes matter in ways that enhance order through the creation of life. The unused energy is returned as waste heat to the cold sink of space. According to one estimate, the total energy flow used by Earth creatures is about 10 trillion calories per second. But this is only about a tenth of one percent of the total solar energy reaching the surface of the planet. Feinberg and Shapiro speculated that the real lesson of alien biology may turn out to be the ways it has adapted to take advantage of every crumb of available energy—in other words, that the competent use of energy, and not intellectual power, will be the most remarkable quality of extraterrestrials, at least from the human perspective.

Solar radiation is only one of many kinds of energy flow. Other usable types of energy might be the electron-stripping action of cosmic and x-rays on atoms, the gravitational potential of a falling body, and the radiation slowly released by a lump of uranium. The recipe for life offered by Feinberg and Shapiro is dazzling-

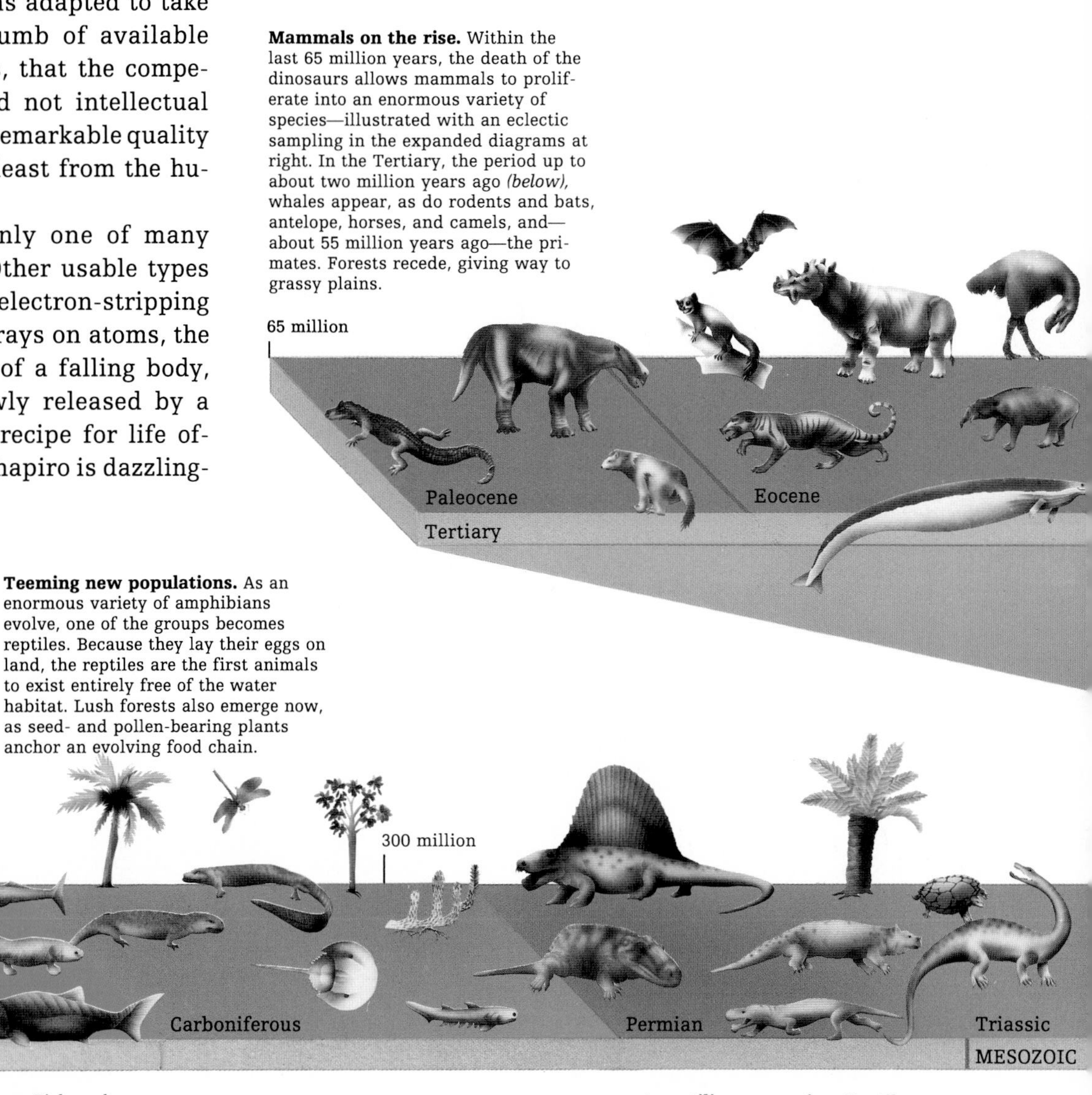

Mammals on the rise. Within the last 65 million years, the death of the dinosaurs allows mammals to proliferate into an enormous variety of species—illustrated with an eclectic sampling in the expanded diagrams at right. In the Tertiary, the period up to about two million years ago *(below),* whales appear, as do rodents and bats, antelope, horses, and camels, and—about 55 million years ago—the primates. Forests recede, giving way to grassy plains.

Teeming new populations. As an enormous variety of amphibians evolve, one of the groups becomes reptiles. Because they lay their eggs on land, the reptiles are the first animals to exist entirely free of the water habitat. Lush forests also emerge now, as seed- and pollen-bearing plants anchor an evolving food chain.

Invaders from the sea. Fish evolve to jawed forms such as the placoderms *(above).* Others crawl from the water on lobed fins, beginning the line of amphibians—a class that is at home in water and on land. Meanwhile, fuzzy plants take root at water's edge.

A reptilian expansion. Reptiles spread and diversify, many becoming flesh-eating predators. Some return to the sea, including the nothosaur *(above, right),* a fierce hunter. Others evolve into the earliest mammals—hairy animals that nurse their young.

Primates' rapid evolution. Hominids advance to become tool-wielding *Homo habilis (below, right), Homo erectus,* and finally—about 100,000 years ago—modern *Homo sapiens.* Several ice ages punctuate the epoch, providing a climate niche that is filled by numerous cold-adapted species, including the mammoths.

Age of humans. The last ice age ends about 10,000 years ago, marking the beginning of the modern epoch, or Holocene. *Homo sapiens* masters agriculture, metallurgy, and other technologies, coming to dominate Earth's extant species—and even to look beyond their home planet.

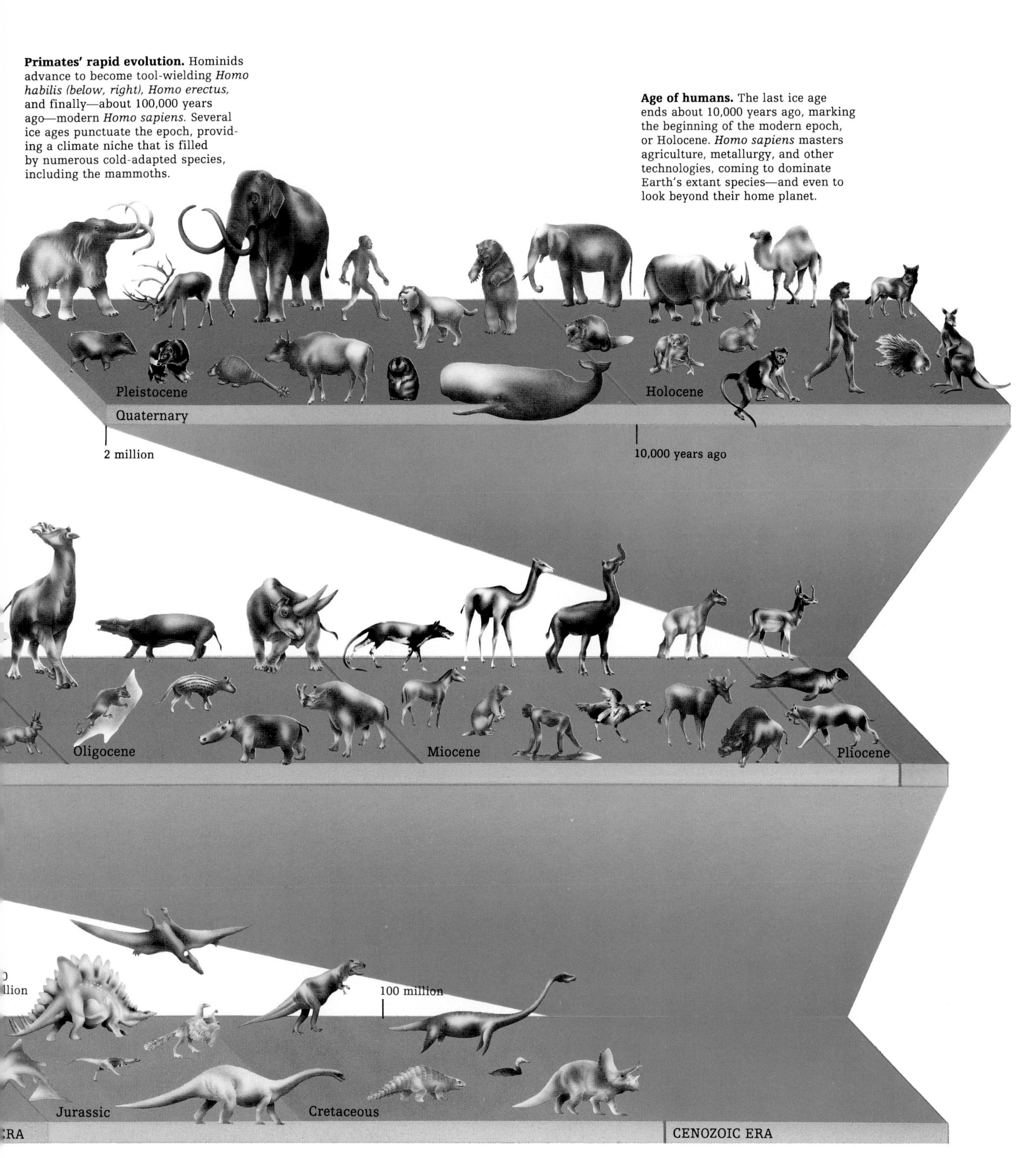

Age of the dinosaur. Reptiles called dinosaurs diverge into many species, some the largest land animals ever known. Meanwhile, flying reptiles and toothed birds take to the air. Forests of conifers, ginkgoes, and cycads flourish, and flowering plants emerge.

The great dying. Dinosaurs continue to dominate the Earth, until a mass extinction that has been variously attributed to a comet crash, continental drift, or climate change. Among the survivors are crocodiles, snakes, turtles, birds, and mammals.

ly flexible: When energy, matter capable of interacting with it, and sufficient time are supplied in one place, "however exotic and unearthly that location may seem, it is where life may be found."

Their theory rests on a redefinition of life broad enough to accommodate all the ways in which all forms of energy can be put to work. They argued that it is a fundamental error to regard the individual as the basic unit of life. In their view, life cannot be seen whole in a cell, or bacterium, a human, or the millions of humans clustered in London or Calcutta. Instead, its basic unit is the realm of life itself—the biosphere. This they defined as "a highly ordered system" in which order is maintained or gradually increased through the interaction of energy with its environment.

The rich vitality of Earth is merely one of many possible expressions of a biosphere in action. Another biosphere, they conjectured, might have no individuals but be instead a single living entity. Alternatively, it might be composed of individuals as closely bound as the cells in the human body. Or, in contrast to Earth, it might favor relatively few species and little difference between individuals.

These ideas about life remain very much a minority view. But Feinberg and Shapiro did not shy away from pressing their scheme to its limits. They argued for the possibility of life anywhere matter can use energy to create order. Conceivably, life could evolve in lakes of frigid ammonia or oceans of oil. Feinberg and Shapiro speculated about silicate creatures living in molten rock, plasma life forms swimming inside stars, radiant organisms inhabiting interstellar dust clouds. Most exobiologists would be disinclined to go that far, but if nothing else, Feinberg and Shapiro, like the authors of the Gaia theory, have served the biological sciences by their wideness of focus: Life may be resourceful in ways we have never imagined.

ON THE EDGE OF EXISTENCE

Zoologists, microbiologists, and other scientists in recent decades have found that life on Earth is nowhere near as frail and particular as once was believed. From the subfreezing Antarctic desert to the sulfurous waters of a hot spring, terrestrial organisms grow and prosper in environments that, by conventional standards, seem almost as inimical as those existing on other planets and moons of the Solar System.

Miles beneath the ocean waves, inhabitants of the abyss prove that utter darkness—a death knell to most plants and animals on the planet's surface and a condition that may obtain, for example, under the ice of Jupiter's moon Europa—is itself no bar to life. Microbe colonies in the core of dry, cold rocks remain an argument for the possibility of similar life on Mars. And species of salt-loving bacteria teach scientists that life's chemical tolerances have more flexible limits than might be indicated by the requirements of more common flora and fauna.

Only one constraint remains: liquid water. No earthly life is known to exist entirely without it; even organisms that seem to subsist on ice or water vapor actually have access to minimal amounts of liquid water. For now, exobiologists must assume that on other worlds, as on Earth, life and water go hand in hand.

BATTLING COLD AND DRYNESS

Frigid Antarctica is the least hospitable continent on Earth and the closest terrestrial analog to the planet Mars *(far right).* Exobiologists are thus very encouraged to find that it is home to thriving populations of lichen, algae, fungi, bacteria, and insects. These creatures, along with oddities such as the tardigrade *(below, right),* suggest that extreme cold and a lack of

Preserved by a kind of biological antifreeze, the Antarctic snow flea *(left)* leaps through air too cold for flight with an action that gives rise to its other common name: springtail. The flea bends a forked appendage beneath its abdomen, then releases it. Like a freed spring, the fork hurls its owner up to pursue food or avoid predators.

Massed snow algae *(left)* stain a patch of Antarctic snow bright pink. The pigment is believed to shield the algae from solar ultraviolet rays, and a chemical component keeps them from freezing. The algae absorb minute amounts of nutrients, but they are primarily photosynthesizers, regulating their solar intake by adjusting their depth in the snow.

Slow-growing lichen, algae, fungi, and bacteria abound in the Antarctic desert. All live under the surface of porous rocks, like the sliced sandstone below, a sample from the Ross Desert. Penetrating sunlight provides energy for photosynthesis, while the insulating rock wards off the worst of the cold.

water need not preclude life on Mars-like planets.

The Martian comparison is apt. Water ice covers most of Antarctica just as carbon dioxide ice and water ice crown the Martian poles year-round. Antarctica's arid plains suffer temperatures as low as minus 125 degrees Fahrenheit, minus 165 with wind chill. Temperatures in the Martian drylands are as cold or colder, the atmosphere is even thinner, and there is little free oxygen and no liquid water.

While Mars is arid now, its surface is marked with signs of water: Deep channels and teardrop-shaped islands indicate that a torrential flow, equivalent to 10,000 Amazon rivers, once coursed through the red terrain. The planet's atmosphere was probably denser then, and conditions may have been just temperate enough for life, perhaps on the Antarctic model. Some exobiologists believe that the fossilized remains of microscopic Martian life may await discovery in now-dry lake beds or rocks. A speculative few hold out hope for living microbes at the borders of the Martian permafrost.

Magnified below more than 1,400 times, the tardigrade is an earthly anomaly. Found in moist mud or moss, the creature is 85 percent water but can reduce to as little as 2 percent. A dried, dormant tardigrade can survive heating to 240 degrees and cooling to minus 400. Returned to room temperature and revived with water, it can walk away unharmed.

SURVIVAL UNDER PRESSURE

The specialized creatures of Earth's deep sea exist at almost unthinkable pressures. Fish and shrimp have been sighted 35,000 feet down in a Pacific ocean trench, swimming easily in spite of a water pressure of eight tons per square inch, or more than a thousand times the atmospheric pressure at sea level.

Deep-water life forms are often striking in appearance because of their adaptations to the lack of light.

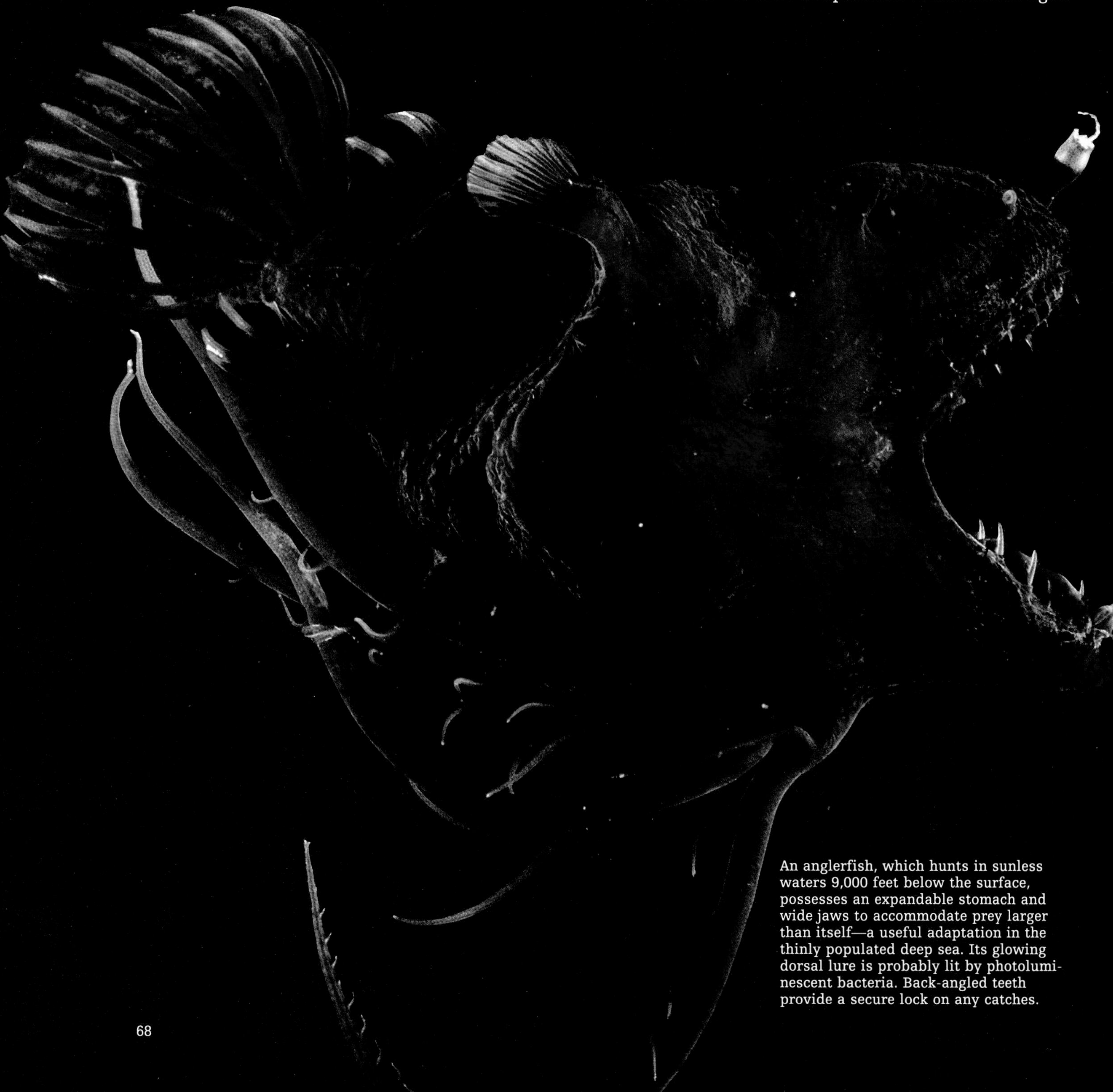

An anglerfish, which hunts in sunless waters 9,000 feet below the surface, possesses an expandable stomach and wide jaws to accommodate prey larger than itself—a useful adaptation in the thinly populated deep sea. Its glowing dorsal lure is probably lit by photoluminescent bacteria. Back-angled teeth provide a secure lock on any catches.

But the adaptation that is most crucial to their survival occurs at the molecular level. If surface creatures were suddenly transported to the deeps, the very molecules making up their cells would be so distorted by the enormous pressure that biochemical processes would cease.

Deep-sea fish, in contrast, have developed molecules that form unusually rigid, pressure-resistant proteins, thus allowing normal enzymatic reactions to take place. Adaptations such as these open the door to the possibility that some sort of creatures might master conditions in the dense atmosphere of Jupiter *(above, right),* where pressures a few hundred miles beneath the cloud tops resemble those of Earth's deep oceans.

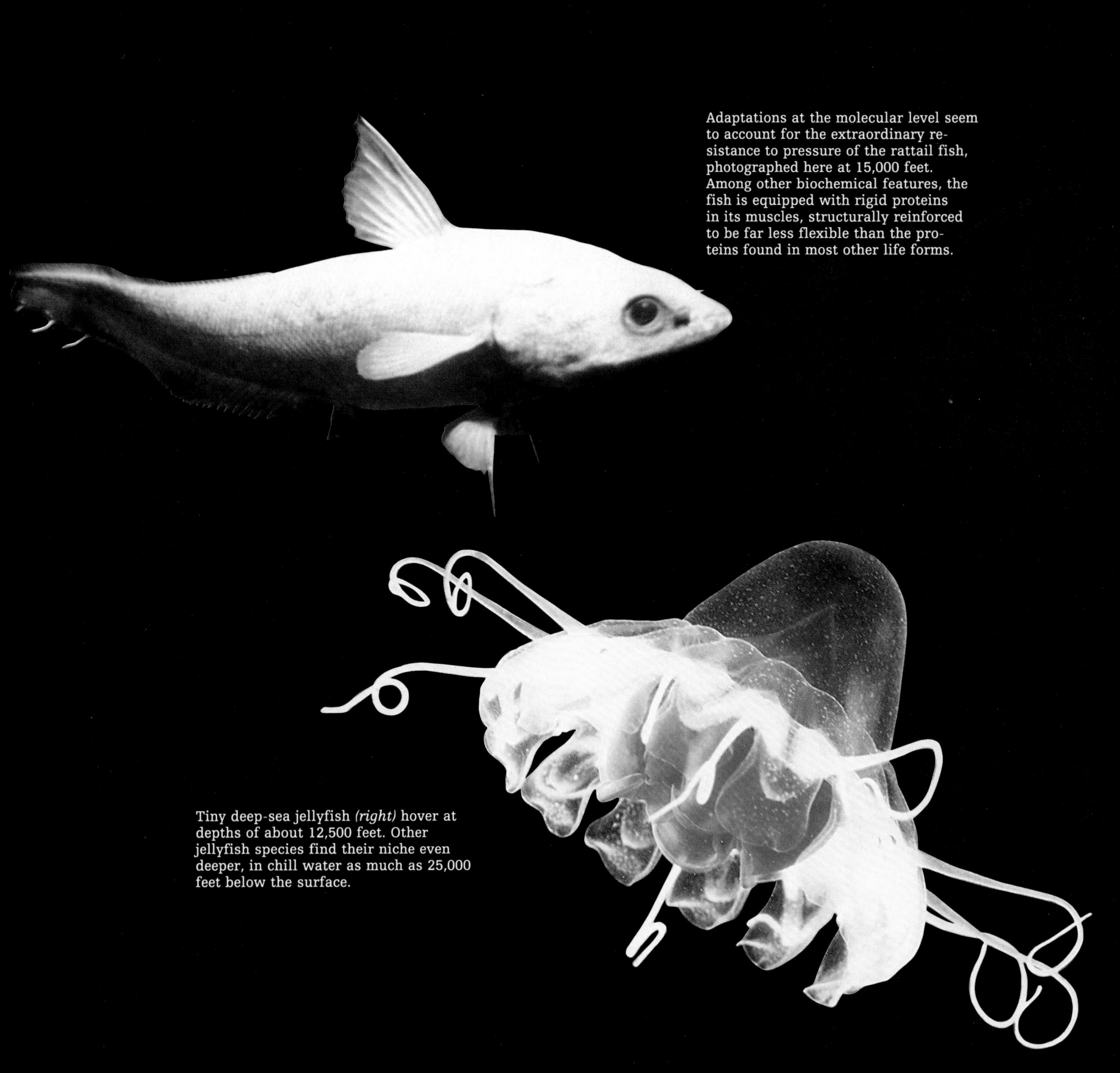

Adaptations at the molecular level seem to account for the extraordinary resistance to pressure of the rattail fish, photographed here at 15,000 feet. Among other biochemical features, the fish is equipped with rigid proteins in its muscles, structurally reinforced to be far less flexible than the proteins found in most other life forms.

Tiny deep-sea jellyfish *(right)* hover at depths of about 12,500 feet. Other jellyfish species find their niche even deeper, in chill water as much as 25,000 feet below the surface.

In the Absence of Sunshine

Biologists once thought that all life required sunlight: If a life form did not directly use solar energy through photosynthesis, it must be part of a food chain beginning with plants that did. Then came the discovery of chemosynthetic communities near pitch-dark hydrothermal vents on the ocean floor. In these ecosystems, organisms derive energy not from sunlight but from minerals seeping out of Earth's core. The find opens up new places to look for extraterrestrial life. On Jupiter's moon Europa *(above, left),* for example, a water-ice crust may cover an ocean as much as thirty miles deep. At the bottom of this sunless realm, the Europan version of hydrothermal vents might fuel analogous chemosynthetic colonies.

The first link in the deep-sea food chain is chemosynthetic bacteria *(inset),* which oxidize hydrogen sulfide from vented waters, releasing energy to make carbohydrates. Some organisms, including small crustaceans, feed directly on the bacteria. Others, such as tube worms *(left),* live with them in symbiosis: With their feathery tentacles, the nine-foot-long worms concentrate sulfides and other minerals from the vents inside their own bodies, where the bacteria wait to go to work.

A Taste for Salt

For most terrestrial life, highly saline water is deadly. Utah's Great Salt Lake, for example, is eight times saltier than the ocean and is totally lacking in fish. But certain species of shrimp, flies, bacteria, and algae thrive in its waters. Related halophilic, or salt-loving, organisms are also found in Israel and Jordan in the even more saline Dead Sea. Still others cluster in shallow ponds where seawater is evaporated to provide salt for human consumption.

Saline water poisons ordinary cells because of a phenomenon called osmotic pressure, created by differences in salt content between the cell interior and the surrounding fluid. Pumping out water in a vain attempt to equalize saline levels on either side of the cell wall, the cell shrivels up. Many halophilic organisms counteract this tendency with microscopic pumping systems that import filtered water as quickly as osmosis extracts it. Like other highly adapted creatures, however, halophiles are trapped in their peculiar niche: Placed in ordinary seawater, with too little osmotic pressure, they overfill from their own pumping, burst, and die.

Halophiles lead some scientists to theorize about life on a salt-dominated world, but few regard the nearest—Jupiter's moon Io *(above, left)*—as a likely spot. Because it is subject to constant resurfacing by sulfurous lava flows, Io is considered too unstable a body to maintain life.

A species of crustacean known as brine shrimp lives and breeds in waters salty enough to kill other crustaceans and any fish. Brine shrimp eggs laid in evaporating pools can remain there in suspended animation for months or more, then hatch within hours when salt water is restored.

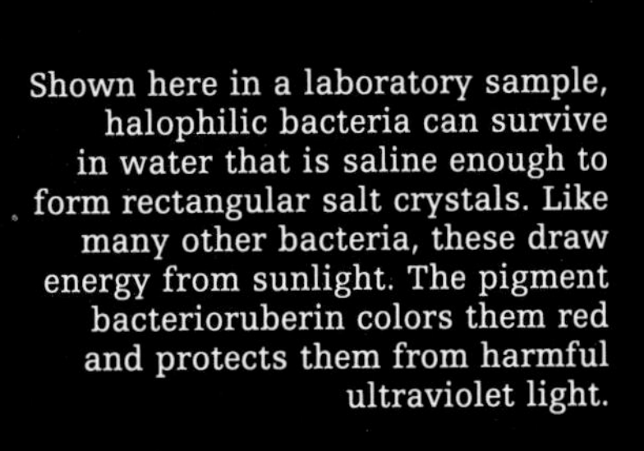

Shown here in a laboratory sample, halophilic bacteria can survive in water that is saline enough to form rectangular salt crystals. Like many other bacteria, these draw energy from sunlight. The pigment bacterioruberin colors them red and protects them from harmful ultraviolet light.

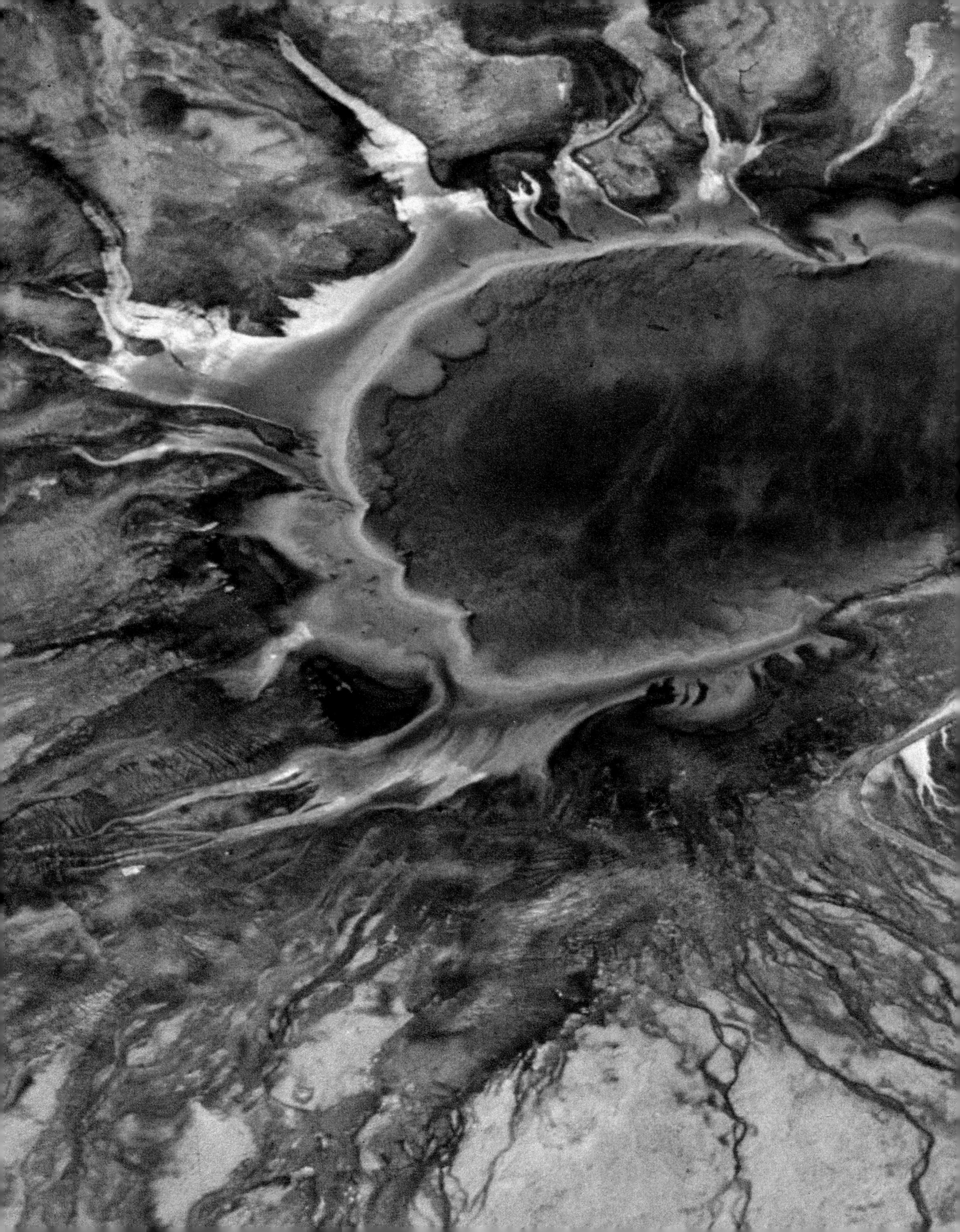

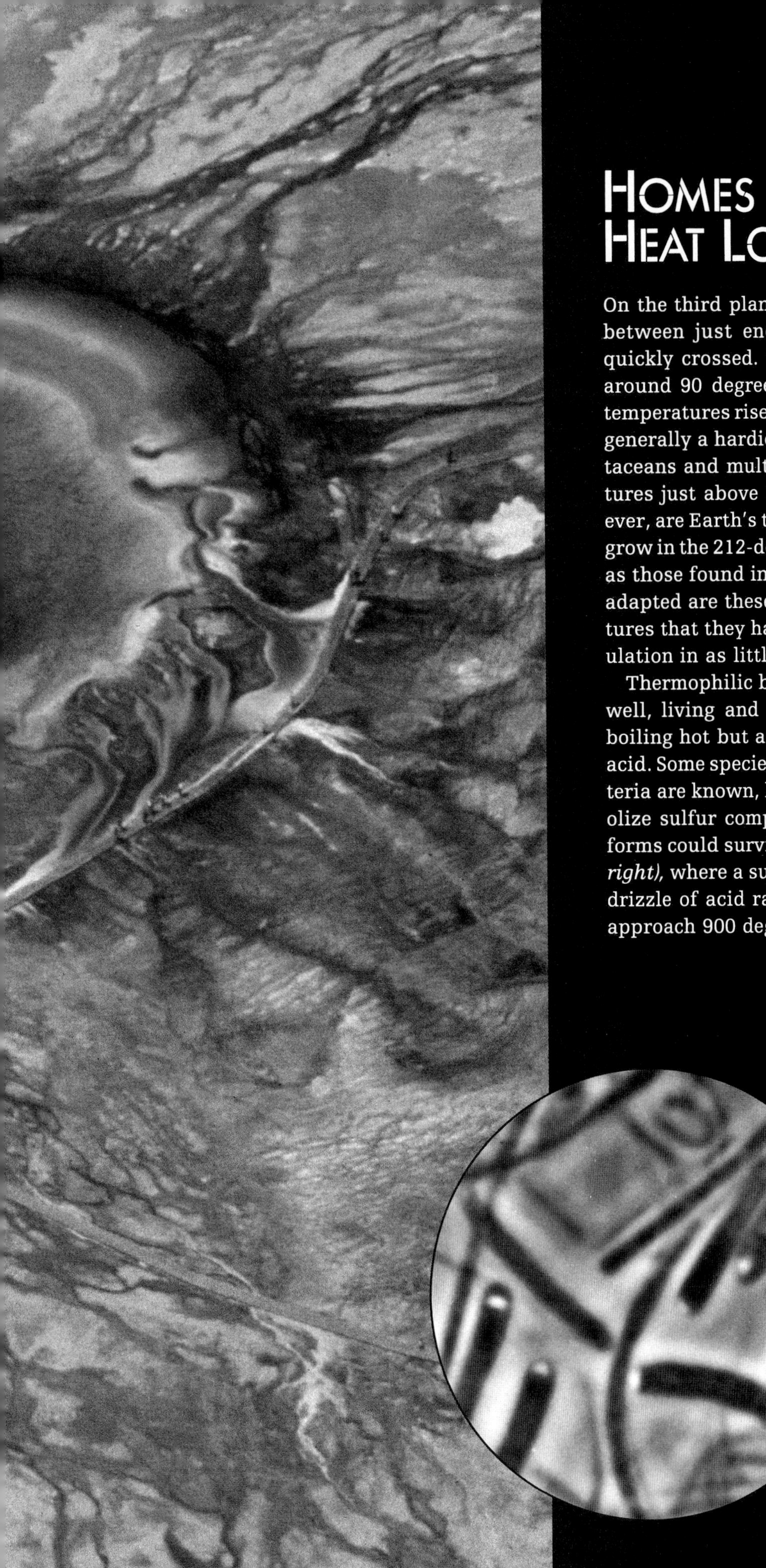

Homes for Heat Lovers

On the third planet from the Sun, the line between just enough heat and too much is quickly crossed. Tropical fish that do best in water around 90 degrees Fahrenheit will cook to death if temperatures rise much above 100 degrees. Insects are generally a hardier family, but they, along with crustaceans and multicellular plants, perish at temperatures just above 120 degrees. Certain bacteria, however, are Earth's thermophiles, or heat lovers, and can grow in the 212-degree waters of thermal springs such as those found in Yellowstone National Park. So well adapted are these thermophiles to scalding temperatures that they have been known to double their population in as little as two hours.

Thermophilic bacteria may tolerate high acidity as well, living and breeding in water that is not only boiling hot but as corrosive as concentrated sulfuric acid. Some species of thermoacidophiles, as these bacteria are known, have developed the ability to metabolize sulfur compounds, suggesting that similar life forms could survive on a planet such as Venus *(above, right),* where a sulfurous cloud layer emits a constant drizzle of acid rain and temperatures at the surface approach 900 degrees.

Colonies of heat-loving bacteria burgeon along the relatively cool edges of Grand Prismatic Spring, an alkaline hot spring in Yellowstone National Park. Two species that live at temperatures just under 160 degrees—*Synechococcus lividus* and *Chloroflexus aurantiacus (inset)*—help tint the pool's perimeter orange and green.

3/The Sun's Neighborhood

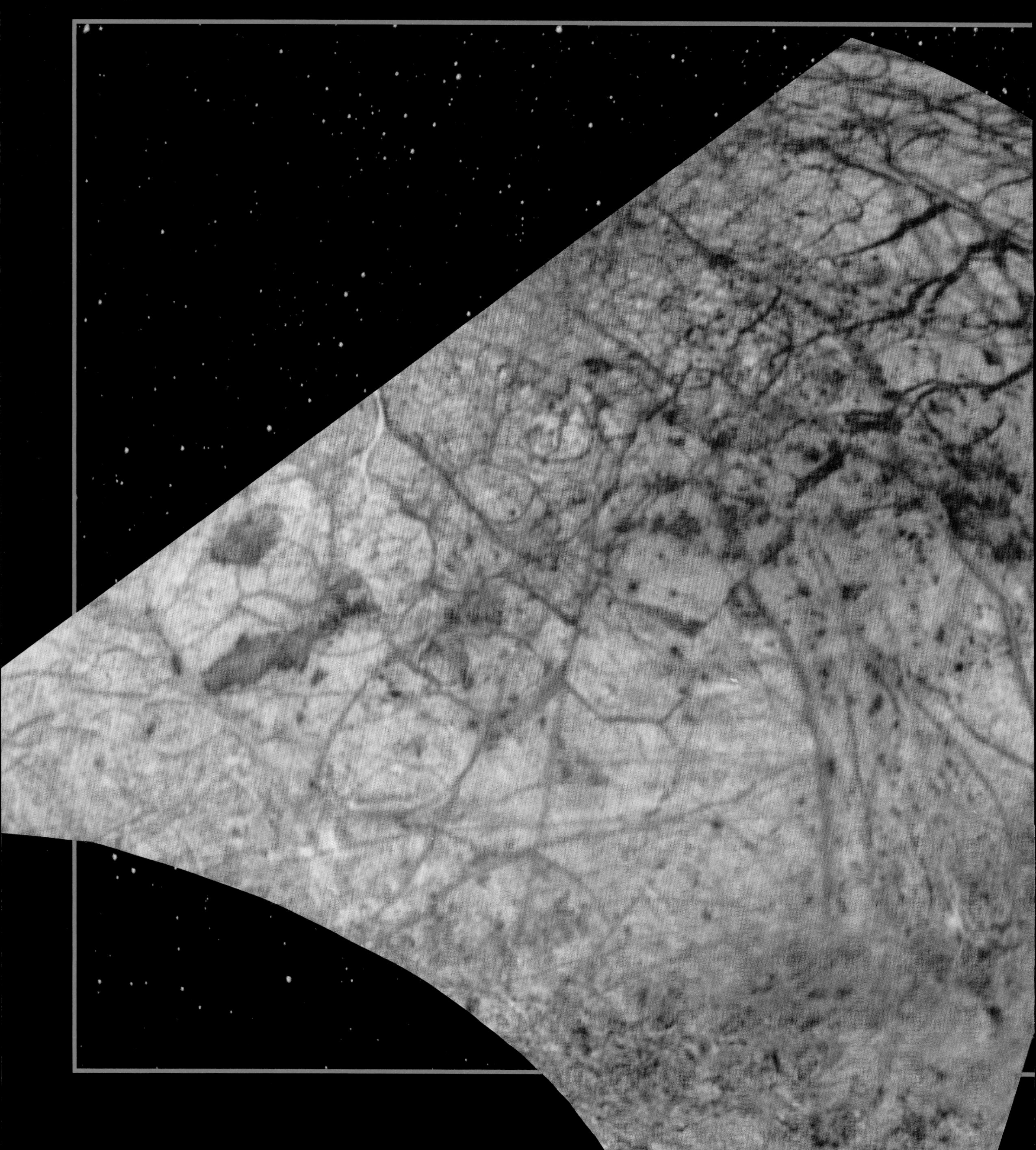

A myriad of cracks, veining the thick surface ice of Europa, Jupiter's fourth-largest moon, may allow sunlight to stir biological activity in an underlying ocean.

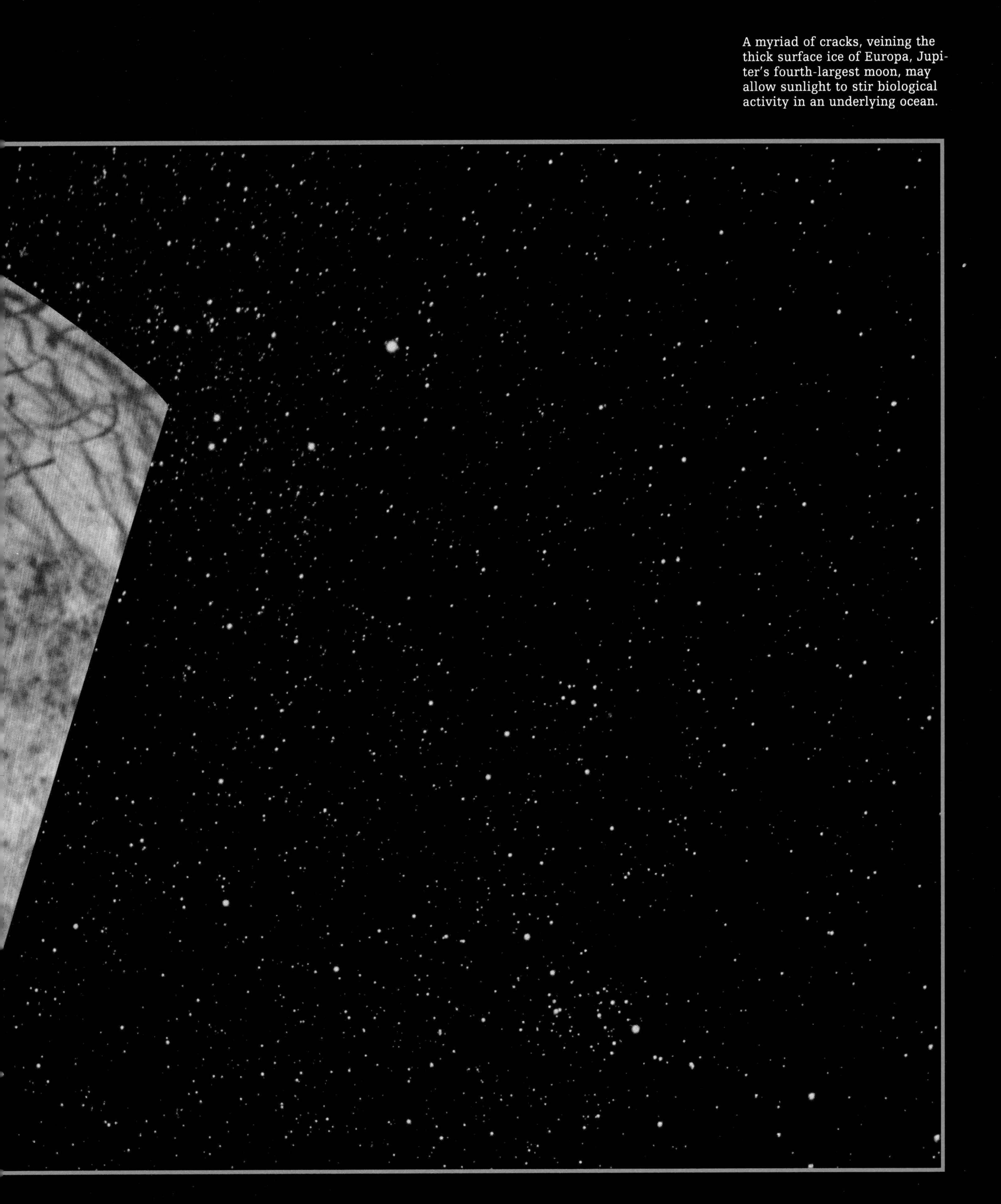

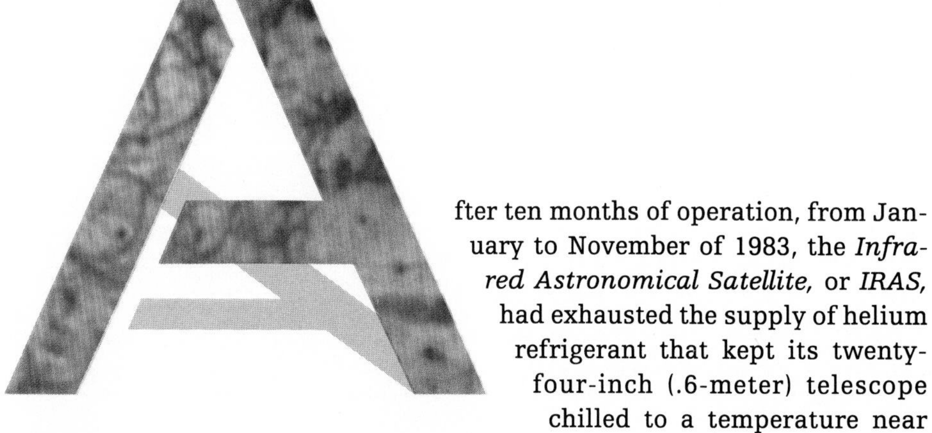

After ten months of operation, from January to November of 1983, the *Infrared Astronomical Satellite,* or *IRAS,* had exhausted the supply of helium refrigerant that kept its twenty-four-inch (.6-meter) telescope chilled to a temperature near absolute zero, or minus 460 degrees Fahrenheit. As the instrument warmed up, its usefulness to ground-based astronomers vanished. *IRAS* was effectively dead. But the little satellite—the joint enterprise of the United States, the United Kingdom, and the Netherlands—had surpassed all expectations its creators may have had, surviving three months longer than its intended life span and returning a wealth of data that would keep astronomers occupied for several years.

The satellite's data was of particular interest to life searchers because infrared—long-wavelength radiation beyond the visible red end of the electromagnetic spectrum—is perhaps the most ubiquitous form of energy in the universe. Anything warmer than absolute zero radiates some infrared energy, and certain molecular combinations such as water, methane, and ammonia, which are vital components in the chemistry of terrestrial life, are easily detected at this wavelength. However, when it comes to observing celestial objects in the infrared, earthbound astronomers are severely handicapped for two reasons. First, they must contend with interference from virtually every warm thing around, including the infrared telescope itself. Second (and the reason *IRAS* was so valuable), only certain wavelengths of infrared can pass unimpeded from space through water vapor in the Earth's atmosphere.

Operating above that atmospheric shield, *IRAS* was able to reveal a new and previously invisible universe. Among other things, for example, the satellite found that giant molecular clouds—tenuous agglomerations of the chemical building blocks of earthly proteins, first detected by radio astronomers in the 1950s—hid warm lumps of matter that seemed to be stars in the making. Another piece of news from the infrared satellite was especially tantalizing: A handful of stars in various parts of the sky—Vega, Epsilon Eridani, Alpha Piscis Austrinus, and Beta Pictoris—seemed to radiate more light at wavelengths in the infrared range of the spectrum than they should have, considering their size, age, and type. Since very hot, ener-

An image generated from infrared data collected by Dana Backman of the National Optical Astronomy Observatories shows a suspected brown dwarf *(opposite)* orbiting Gliese 803, a star in the constellation Microscopium. Brown dwarfs, many times larger than Jupiter but too small to be stars, are theoretical objects whose existence has yet to be proven conclusively. This one cannot be confirmed until astronomers learn more about the object's temperature.

getic objects such as stars are expected to radiate most of their energy at much shorter wavelengths, from visible light up to gamma rays, astronomers took this excess stellar infrared to be the signature of some other infrared-radiating object nearby. They speculated that the stars in question were each surrounded by a shell or disk of fine dust particles, which would offer a substantial amount of radiating surface area for their mass. However, images constructed from the *IRAS* data simply could not resolve the source of the excess radiation.

The uncertainty left many in the astronomy community on tenterhooks. The formation of planets around their stars is believed to begin with the accretion through gravity of a circumstellar disk of material, but estimates of how often this event occurs in the galaxy range widely *(pages 90-93)*. If the excess infrared proved to be emanating from a disk, it would be the first significant piece of evidence that other solar systems exist.

In April 1984, a few months after the *IRAS* findings had been made public, two members of the imaging team for NASA'S Voyager mission to the outer planets traveled to the Carnegie Institution's Las Campanas Observatory in the mountains of northern Chile. Bradford Smith of the Jet Propulsion Laboratory and Richard Terrile of the University of Arizona in Tucson had helped obtain Voyager's extraordinary closeup views of Jupiter, Saturn, and their moons. At the moment, they were between flybys: Voyager's rendezvous with Saturn had ended in 1981, and its encounter with Uranus was not due until January 1986. Uranus was their primary objective on this observing run, but both men were intrigued by the *IRAS* reports of possible circumstellar disks. The prospect of verifying such a disk—and finding a whole new solar system—was irresistible.

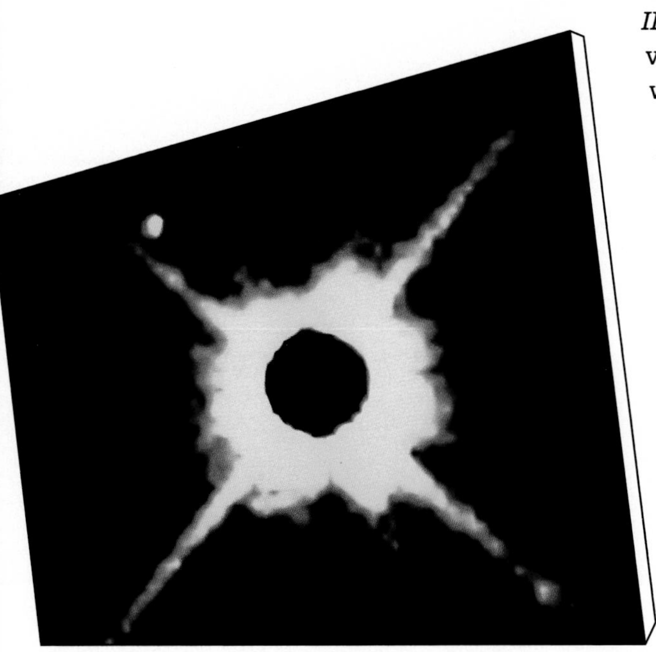

Beta Pictoris, one of the stars mentioned in the *IRAS* report, was within their field of view from the Las Campanas facility. With twice the mass and seven times the luminosity of Earth's own Sun, "Beta Pic," as astronomers familiarly call it, lay in Pictor, an obscure constellation that is best viewed from the Southern Hemisphere. The two astronomers decided to try to take the first visible-light image of it.

In order to discern something that would ordinarily be hidden in the glare of the star's own light, the pair employed a coronograph, an instrument regularly used in observing the Sun's tenuous upper atmosphere, or corona. They attached the device to the observatory's powerful 100-inch (2.5-meter) telescope, along with a camera equipped with a superchilled electronic light detector.

After four nights of observing Beta Pic in the far-red part of the visible spectrum, Smith and Terrile took their data tapes back to computers in Pasadena and Tucson for processing. "It was really unbelievable," Terrile recalled later.

A disk of dust particles circling the star Beta Pictoris—shown here in a computer-generated image based on data gathered by University of Arizona astronomer Bradford Smith *(below)*—hints at the possibility of planet formation. The false colors represent different intensities of light reflected from the disk, which measures about 100 billion miles across. (Beta Pictoris itself is partially blotted out at center to reduce glare.)

"We put together all the images, and all of a sudden there was something we couldn't explain."

The final computer model revealed the edge of a circumstellar disk extending some 100 billion miles outward from Beta Pictoris—about thirty times the average distance of Pluto from the Sun. Highly flattened and thickening gradually with distance from the star, the disk seemed to be composed of tiny particles, possibly small silicate grains, carbon compounds, and water ice. Smith and Terrile noted that the marked flattening of the disk indicated that the system had not been disturbed by the gravitational passage of other cosmic bodies in its lifetime, suggesting that it might be of relatively recent formation. The disk appeared not to extend inward all the way to the star but to be separated from it by 1.5 billion miles, a void that, in the Solar System, would correspond to the radius of the orbit of Uranus.

Although no larger bodies were detected, the two astronomers theorized that the granular material that once filled this gap might have accreted into planets. "It is tempting to speculate," they wrote in their report, "that we may be witnessing ongoing planetary formation around Beta Pictoris or, alternatively, that we are viewing the distant detritus of an already extant but young system of planets."

Beta Pictoris proved to be only the beginning. Three years later, in 1987, a lanky thirty-two-year-old astronomer by the name of Dana Backman, from the National Optical Astronomy Observatories in Tucson, Arizona, completed a study based on data from the *Infrared Astronomical Satellite.* The study looked at nearly fifty star systems within about seventeen light-years of the Sun. Six of them showed the telltale excess of infrared radiation. By the end of another year, Backman had extended the survey out to a radius of sixty-five light-years. Of the 134 star systems examined in this greater solar neighborhood, as many as twenty-five had excess emissions indicating that the stars were encircled by disks or clouds of particles. Further examination of the data suggested that two of them displayed a gap between disk and star similar to the one Smith and Terrile had seen around Beta Pictoris.

INTERESTING AND WONDERFUL THINGS

With as many as one in five or six stars in the Sun's neighborhood apparently haloed by planet-forming material, the next logical step is to determine whether planets have actually formed. Given the distances between Earth and these potential solar systems and the fact that planets are small, dim, and cold compared with their parent stars, executing that step is clearly a case of easier said than done. Undaunted, astronomers such as Dana Backman have harnessed innovative technologies and techniques both direct and indirect for the pursuit of their elusive quarry.

Late in 1987, in the course of extending the National Optical Astronomy Observatories infrared survey, the young scientist was conducting an observing run at Cerro Tololo Interamerican Observatory in northern Chile.

Observatories strung like gleaming white beads along the barren crest of Mauna Kea house some of the powerful instruments used to hunt for signs of extrasolar planetary systems. Well below the 13,800-foot summit, a tropical inversion cloud layer prevents moist sea air from rising into the upper atmosphere. The extremely dry summit air, coupled with minimal light pollution, makes the extinct Hawaiian volcano perhaps the world's best site for astronomical viewing.

Thus far, the run had been disappointing. Four of his five observing nights had been ruined by the wet weather of the southern spring, with clouds shrouding the 7,000-foot-high observatory. On this final night, however, the sky had cleared.

Ordinarily, Backman's search for what he called "interesting and wonderful things" would have been severely limited: Cerro Tololo's sixty-inch optical reflector was at best a moderate-performance telescope by modern standards. But a few months before, a new device had transformed it into one of the most powerful infrared instruments in the world.

Attached to the telescope was a camera that was kept chilled by liquid nitrogen and helium, and inside the camera, technicians had mounted one of a new family of electronic sensors. Called an infrared detector array, the device consisted of some 4,000 sensors arranged on a tiny grid. Not only did it drastically reduce the amount of telescope time normally required to gather sufficient infrared data, the infrared detector array represented a hundredfold improvement in sensitivity over old-style sensors, which had only a single element. The camera used by Backman also had a star-blocking feature similar to the one used by astronomers Smith and Terrile for their observations of Beta Pictoris.

Backman's targets were Gliese 803 and Gliese 879 (also known as AU Microscopium and TW Piscis Austrinus for their respective constellations). Two black-and-white video screens guided the observations, one connected to a video camera used to point the telescope, the other to the computer receiving signals from the infrared detector array. As the computer churned away on incoming data, the star fields on the computer monitor underwent subtle transformations. Near Gliese 803 and again near Gliese 879, a pinpoint of light popped out of the background. Backman was seeing the infrared glow of previously uncharted companions.

The two dim points of infrared light might have been low-mass stars in orbit around Gliese 803 and 879. They might also have been more distant background objects. However, Backman believed that they were strong candidates for a theoretical object that was neither quite star nor quite planet, known as a brown dwarf.

FAILED STARS

This class of objects is, in effect, the product of theoretical necessity. Calculations of how strong a gravitational field would need to be to hold clustered galaxies together indicate that some ten times more mass is needed than has actually been detected. The hypothetical brown dwarf, too dark to be seen, is one contender for some of this missing material. As imagined, brown dwarfs would be intermediate in mass between the largest planets in the Solar System and the smallest possible stars, which require at least eight percent of the Sun's mass to ignite their thermonuclear fusion engines. Such substellar objects could have masses up to eighty times that of Jupiter and would retain enough primordial heat to

have a surface temperature of about 2,000 degrees Fahrenheit, just right to make them bright at infrared wavelengths. Some might exist alone as failed stars in interstellar space; others could be in orbit around more successful companions.

Backman's candidates were not the first to be put forth, nor would they be the last, thanks in part to an instrument that seemed custom-made for finding brown dwarfs: the 120-inch (three-meter) Infrared Telescope Facility, or IRTF. Equipped with a new infrared detector array, the telescope had been built by NASA on the crest of Mauna Kea, an extinct volcano on the island of Hawaii. At nearly 14,000 feet above sea level, Mauna Kea rises above much of the atmosphere's water vapor and turbulence. Only outer space—with no atmosphere to cause twinkling and blurring of celestial light—offers better "seeing," as astronomers term observing conditions.

From this vantage point, in August of 1987, Benjamin Zuckerman of the University of California at Los Angeles and Eric E. Becklin of the University of Hawaii detected a powerful infrared emission in the vicinity of Giclas 29-38, a white dwarf star located in the constellation Pisces. They deter-

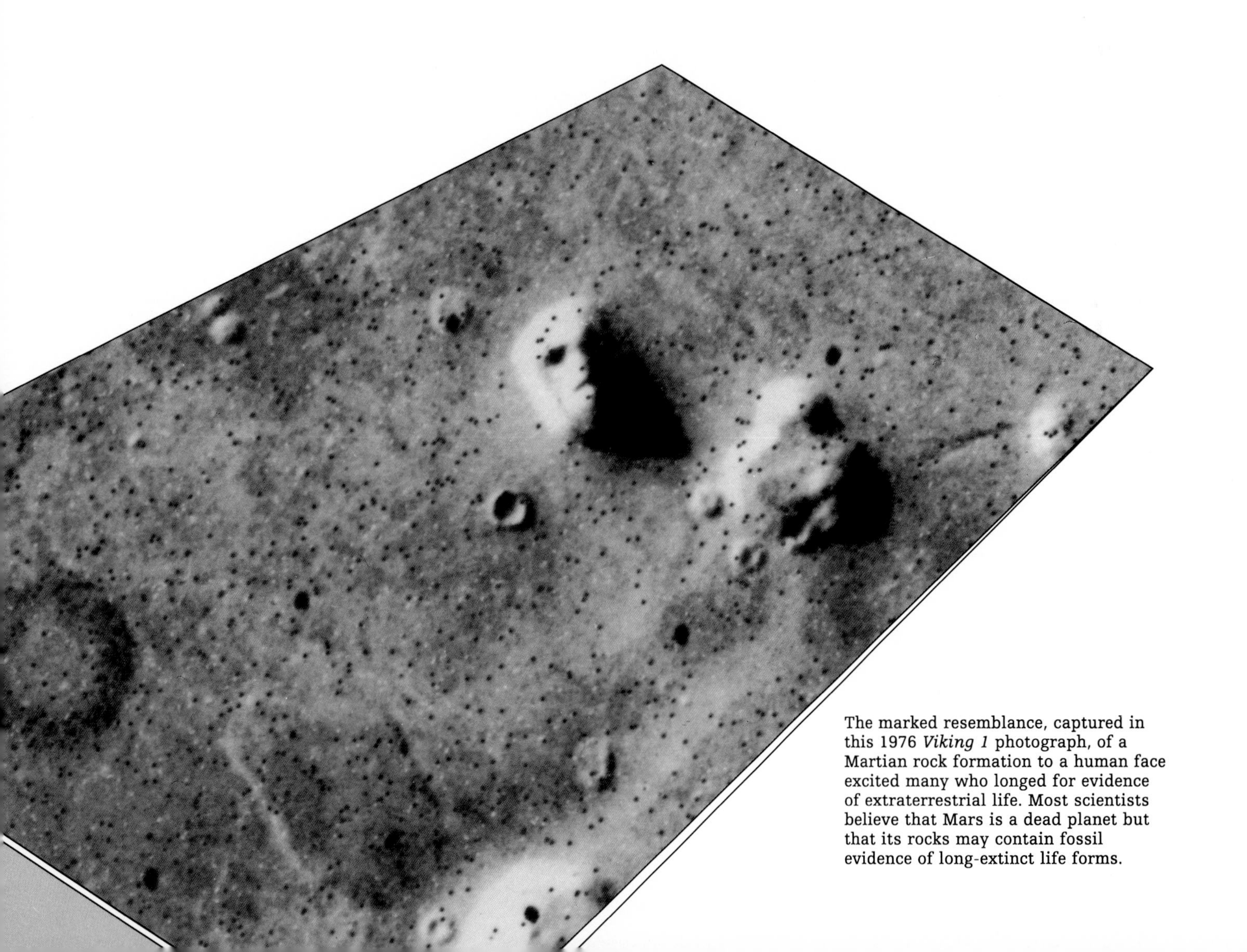

The marked resemblance, captured in this 1976 *Viking 1* photograph, of a Martian rock formation to a human face excited many who longed for evidence of extraterrestrial life. Most scientists believe that Mars is a dead planet but that its rocks may contain fossil evidence of long-extinct life forms.

mined that the characteristics of the signal resembled those calculated for brown dwarfs. By their reckoning, the object would have a radius about 15 percent that of the Sun and would contain between 6 and 8 percent of a solar mass—in other words, about sixty to eighty times the mass of Jupiter.

The discovery launched a seeming run on the rare species: In mid-1988, a team of researchers from the University of Rochester and Cornell analyzed earlier IRTF observations and reported another possible brown dwarf. Zuckerman and Becklin looked at their 1987 data again and found yet another candidate, and late in 1988 Zuckerman and Becklin reported a "very convincing" case for a brown-dwarf-like object near Giclas GD 165 in the constellation Boötes.

LOWER LIMITS

Abundant as they now seemed to be, however, brown dwarfs are as elusive as snipe. As astronomers continually reexamine the evidence for every claim, serious questions arise. For example, Zuckerman and Becklin's finding of excess infrared radiation around Giclas 29-38 might, according to one countertheory, be an indication of a circumstellar disk rather than a substellar body. Clearly the jury is still out. It will probably remain so for the foreseeable future since brown dwarfs and super-Jupiters seem to constitute the lower limit of what present technology can detect with direct methods such as those of ground-based infrared astronomy. Astronomers hunting for these invisible worlds have also taken an indirect approach. The trick is not to look for the planets themselves but to seek evidence of their gravitational effects on their central star. Contrary to appearances, a star such as the Sun actually forms a binary pair with each of its planets individually: In the Solar System, planet and Sun orbit one another, circling a common center of gravity known as the barycenter. The Sun thus endures a series of gravitational displacements away from each planet, with Jupiter exerting by far the largest influence.

These stellar motions—even those caused by planets as massive as

Jupiter—are vanishingly small. To detect them in distant solar systems, astronomers study stellar spectra, an array of lines corresponding to various wavelengths of light. Spectra are produced when starlight passes through a device called a spectrograph. Depending on a star's motion relative to Earth, a phenomenon known as the Doppler effect can slightly shift the lines of the star's spectrum—toward the shorter wavelengths at the blue end of the spectrum if the star is moving in the direction of Earth, and toward the longer, red wavelengths if it is moving away. When the star is moving parallel to the Earth's direction of motion, no Doppler shift occurs, and the lines in the star's spectrum correspond to those produced by a light source at rest. If a pattern of shifts repeats periodically, the star may be in a mutual orbit with an unseen companion.

The technique has long been used to identify stars in binary or multiple systems, orbited by another star or stars too dim to be seen. With objects of stellar mass, the changes in velocity are relatively conspicuous: a thousand or so miles per hour in binaries of equal mass, for instance. Something with only the mass of a planet, however, would produce velocity changes a hundred times smaller.

LOOKING FOR SUBTLE MOTIONS

In the early 1980s, a team of Canadian astronomers began a long-term project to study these subtle motions. Their instrument was the 144-inch (3.6-meter) Canada-France-Hawaii telescope, one of the ten largest telescopes in existence and, thanks to its location atop Mauna Kea, one of the most capable. To zero in on the minuscule variations in stellar velocity, astronomers Bruce Campbell, Gordon Walker, and Stephenson Yang inserted an innovative filter on the telescope between the incoming light and the spectrograph: a yard-long tube filled with hydrogen fluoride, a colorless gas so corrosive it must be sealed in its vial with windows of sapphire. Once the filter was in place, the astronomers could obtain two superimposed spectra at the same time, one from the star and the other from the gas, which would absorb the star's light in a predictable way. By comparing the star's spectrum against the characteristic, sharply defined spectral pattern produced by the hydrogen fluoride, the team could spot wavelength shifts corresponding to velocity changes as small as thirty miles per hour—sensitive enough to detect the slight tugging of a body at least the size of Jupiter.

Gerard P. Kuiper, shown here in front of the University of Chicago's Yerkes Observatory, discovered an atmosphere around Saturn's moon Titan while working at McDonald Observatory in Texas in 1944. The find helped revive serious scientific interest in the search for extraterrestrial life in the Solar System.

For a total of nearly eight years, Campbell and his colleagues used this modified spectrographic technique to make painstaking observations of eighteen stars, storing their data on reels of magnetic tape. At the conclusion of each observing run, Campbell plugged the new information into the existing data bank, then ran the computer program to reveal the movements of each star under their watch. In August 1988, Campbell presented his findings at a meeting in Baltimore of the International Astronomical Union. Of the stars they had examined over the years, half had displayed the minute varia-

ALTERNATIVES TO CARBON

An important issue for exobiologists is the viability of biochemistries based on something other than carbon, the terrestrial standard. A leading contender for an alternative is silicon, carbon's closest chemical cousin in the periodic table of elements. It makes up 28 percent of the Earth's crust and accounts for some 7 percent of the matter in the Solar System.

Like carbon, silicon is a versatile element that can knit together complex chemical chains *(right).* As silicon dioxide, or quartz, it is the chief constituent in glass, high-tech ceramics, and abrasives. It also bonds with oxygen in long molecular chains called silicones. These come in several varieties; depending on the organic groups attached to the silicon atoms, they may be fluid, resinous, stable at high temperatures, or water-repellent. Silicon-oxygen bonds are so strong, however, that once formed, silicones tend to be self-contained, rarely reacting with other compounds, which bodes poorly for building the more elaborate structures that lead to living things.

Terrestrial life does use silicon in some ways. Simple sea creatures called diatoms form outer skeletons from the element. Small protozoa known as radiolarians build a body-supporting lattice from it. On land, horse-tailed rushes reinforce their slender stalks with a silicon framework. But in all these cases, silicon's role is superficial: Silicon supplies an outer structure, but the actual living tissue in diatoms, radiolarians, and rushes is carbon based.

Researchers have also looked at the possibility of life based on other elements, including nitrogen and phosphorus, both critical to life on Earth. But such materials seem less promising even than silicon, forming fewer complex structures and much weaker bonds. More radical yet is a suggestion that extraterrestrial creatures might be made from tungsten, the metal used for filaments in incandescent lights, an idea that has been received with some skepticism. As one biochemist responds to all such hypotheses, "Fine. First show me how it's done."

tions in velocity that seemed to signal the existence of companion planets.

The news was electrifying, but Campbell urged caution. Only one of the stars, Gamma Cephei, had been observed throughout an entire orbital cycle. It showed a regular, cyclic perturbation indicating the presence of a companion with about 1.6 Jupiter masses. The other candidates had yet to be studied through one or more of their apparent planetary years. "The current evidence is tantalizing," said Campbell, "but we need still more to confirm the existence of extrasolar planetary systems. At least we can say we're hot on their trail."

REAL LIMITATIONS

The prospect of discovering that relatively nearby stars have given birth to planets is heady wine to exobiologists hunting the universe for potentially life-bearing worlds. But even if a full-fledged solar system were sighted beyond our own tomorrow, terrestrial scientists would have only the most limited means of studying it. At best, they might look for chemical signs of life in the runic scrawl of the other system's spectral lines. For the time being, scientists interested in any actual investigation of other worlds, whether by robot proxy or human spacefarer, must set their sights on candidates residing in the more immediate neighborhood.

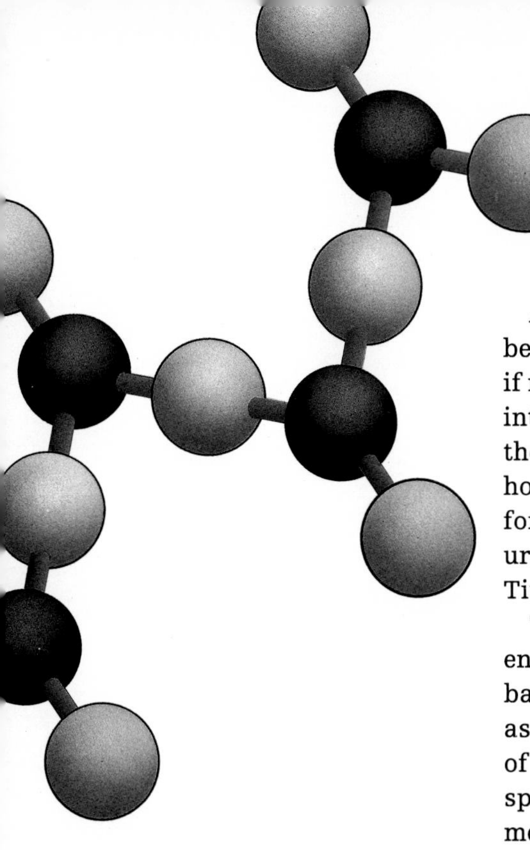

Although the results of Viking's experiments on Mars were ambiguous at best, the next set of missions to the Red Planet may yet find signs of life past, if not life present. Moreover, other members of the Sun's family present some intriguing possibilities. Perhaps of greatest fascination these days is Titan, the largest of Saturn's seventeen moons. Some time in the mid-1990s, NASA hopes to launch a mission to explore both moon and planet. Named Cassini for the seventeenth-century astronomer who discovered a major gap in Saturn's rings, the spacecraft is designed to drop an instrumented probe through Titan's orange canopy of clouds.

The clouds are in fact what had fired the imagination of life-search scientists even before the first closeups of the ringed planet's moon were sent back by *Voyager 1* in November 1980. Nearly four decades earlier, Dutch-born astronomer Gerard Peter Kuiper *(page 84),* who was working at the University of Chicago's McDonald Observatory in Texas at the time, had analyzed Titan's spectrum and determined that it was surrounded by a gaseous envelope of methane. A pioneering scientist widely regarded as the father of infrared astronomy, Kuiper lived long enough to help shape NASA's budding program of planetary exploration but died in 1973, four years before the Voyager spacecraft were launched.

In Voyager's images, the envelope detected by Kuiper was confirmed: The moon's surface remained completely hidden by a shroud a hundred miles or so deep. The probe's instruments indicated that this atmosphere is more than one and a half times as dense as Earth's, resulting in an atmospheric mass ten times greater than that on the third planet. Titan is spared a crushing surface pressure only by its lesser gravity.

As it flew by on its way out of the Solar System, *Voyager 1* beamed back a stream of detail on the chemical makeup of the moon's thick covering. Where Kuiper had thought Titan's atmosphere was mostly methane, Voyager showed it to be nearly 90 percent nitrogen—a figure that compares favorably with the 75 percent nitrogen of Earth's atmosphere. As far as biologists are concerned, Titan's atmosphere is a virtual factory for the production of organic compounds: The portion of the mixture that is not nitrogen is rich with methane, hydrogen, argon, carbon monoxide, carbon dioxide, and organic molecules such as ethane, propane, and hydrogen cyanide, a key intermediate in the synthesis of amino acids. The JPL team that analyzed Voyager's data on the moon did not go so far as to suggest outright that conditions on Saturn's moon were suitable for life. "However," they reported cautiously, "we do note that the chemical processes leading to the formation of organic molecules, precursors of biologically important compounds, appear to have occurred on Titan as well as Earth."

Were Titan endowed with a warmer climate, the moon would almost certainly be a cauldron of such prebiotic chemical activity. In the view of some researchers, however, its surface temperature of minus 289 degrees Fahrenheit precludes the existence of life—or at least life as terrestrial scientists currently recognize it. "While hydrogen cyanide undoubtedly spawns more complicated compounds in the Titanian atmosphere," wrote James B. Pollack, a NASA expert in the study of planetary atmospheres, in 1981, "we do not expect it to produce those capable of giving rise to life. There is a missing chemical link—oxygen—which is scarce because its parent molecule—water—lies inescapably frozen to Titan's frigid surface, largely unable to evaporate and interact with other compounds."

With or without oxygen, however, Titan's surface is undoubtedly a splendid natural laboratory for studying the earliest steps in the formation of organic chemicals. According to Pollack, the surface of Titan could possess the richest hydrocarbon deposits in the entire Solar System. Some scientists envision canyons cut in this hydrocarbon landscape by methane rivers, all beneath a steady drizzle of organic molecules. Other scenarios propose that the Saturnian satellite is covered to a depth of half a mile with a cold ocean of ethane, methane, and nitrogen. Either of these hypotheses, if true, bodes well for prebiotic activity of some sort. The reality of Titan must await the direct probe of the Cassini mission, however, which is intended not only to sample the moon's atmosphere on the way down but to study the composition and topography of its hidden surface.

OTHERWORLDLY OCEANS

If Titan has become a prime focus of exobiological interest in the Solar System, it is by no means the only one. The suggestion that two other satellites may possess oceans has aroused considerable scientific curiosity. In 1979, two years before *Voyager 2* reached Saturn, the probe passed within 125,000 miles of Jupiter's moon Europa. The images it beamed back to Earth revealed a surface reminiscent of cracked Arctic sea ice.

Several years after the Voyager flyby, Ray T. Reynolds, a veteran planetary scientist at NASA's Ames Research Center, proposed Europa as the possible site of "the solar system's second ocean." Although the sea-ice analogy had been intended only metaphorically, Reynolds and his colleagues at Ames cited theoretical evidence suggesting that there was in fact an ocean of liquid water beneath Europa's ice crust. For one thing, the moon showed no signs of craters caused by meteorite impacts, indicating that a continual reworking of the crust had covered up the craters in one way or another. Such crustal kneading could be caused by tidal flexing, the stresses exerted on the moon by its gravitational interactions with Jupiter and the planet's other satellites. Tidal flexing would heat the moon's interior and keep its water liquid. The result, according to Reynolds, would be a liquid-water ocean several tens of miles

VENUS: FROM PARADISE TO HELL

Beneath a perpetual shroud of dense clouds, Venus was for centuries impenetrable to everything but human imagination, which conjured up tropical forests and teeming swamps reminiscent of a primitive Earth. In the late 1940s, one artist came much closer to capturing the truth, depicting the Venusian landscape as a dust bowl broken only by wind-etched rock formations *(opposite).*

Both the lush Garden of Eden and the dust bowl succumbed with the advent of radar astronomy in the 1950s and space probes in the succeeding two decades. First, radar revealed that Venus's surface is cratered like other, presumably barren planets and dotted with volcanoes. But worse was yet to come, as early attempts to approach the planet proved.

The Soviets discovered in the 1960s and 1970s that Venus was a fiendishly difficult subject. Three times, probes named Venera (Russian for "Venus") failed as soon as they neared the planet. The next three malfunctioned as they entered the dense atmosphere. So ill-fated did the mission seem that Roald Z. Sagdeev, then director of Moscow's Space Research Institute, quipped, "People think, why should we always go to Venus? Are we sentenced to it?" Finally, between August 1970 and November 1981, eight probes made it to the surface, each surviving long enough to transmit a smidgeon of information. The longest transmission, by *Venera 13,* lasted 127 minutes.

The reason for the landers' abbreviated lives on the ground became clear as soon as their electronic messages were processed back on Earth. The probes revealed a barren, rock-strewn wasteland broiling at more than 800 degrees Fahrenheit. The thick, mostly carbon dioxide atmosphere exerted a crushing surface pressure some ninety-five times that at sea level on Earth. Through it all, lightning crackled and a steady drizzle of sulfuric acid fell from low clouds. In theory, Venus may have been livable once, but over eons the so-called greenhouse effect—the trapping of outgoing heat by carbon dioxide and water vapor in the atmosphere—transformed the second planet into a lethal inferno.

deep—a volume about that of the terrestrial oceans—covered by a thin, massively fissured shell of ice a few miles thick.

Exobiologists generally believe that the temperature and pressure of such an ocean might foster the emergence of some kind of life. True, the amount of solar energy reaching a moon nearly half a billion miles from the Sun would be minimal, but the meager light entering through cracks in the ice could conceivably sustain very primitive biological activity for limited periods. This hypothesis finds some support in the discovery on Earth of chemosynthesizing creatures that are capable of surviving in darkness one and a half to two miles below the surface of the sea around hot springs erupting from the ocean floor *(pages 68-69).* Life might not have originated under such extreme conditions, Reynolds admits, but assuming that simple forms had somehow come to exist in Europa's oceans, they might well continue to survive and prosper.

Yet another ocean may lie even farther from the Sun's rays, on Triton, Neptune's major moon. According to a team of astronomers at the University of Hawaii, the reddish tinge to the moon's atmosphere suggests the presence

Contact: A Matter of Probabilities

As with any endeavor, a sensible first step in the search for intelligent life outside the Solar System is to figure the odds. Such is the essence of a simple mathematical statement known as the Drake equation *(below).* Conceived by American radio astronomer Frank Drake, the equation was meant to stimulate discussion of life-search issues rather than serve as a rigorous scientific formula. Seven factors are used to derive the number, *N,* of civilizations in the Milky Way that are able and willing to transmit and receive interstellar radio signals, the most likely means of communication between worlds.

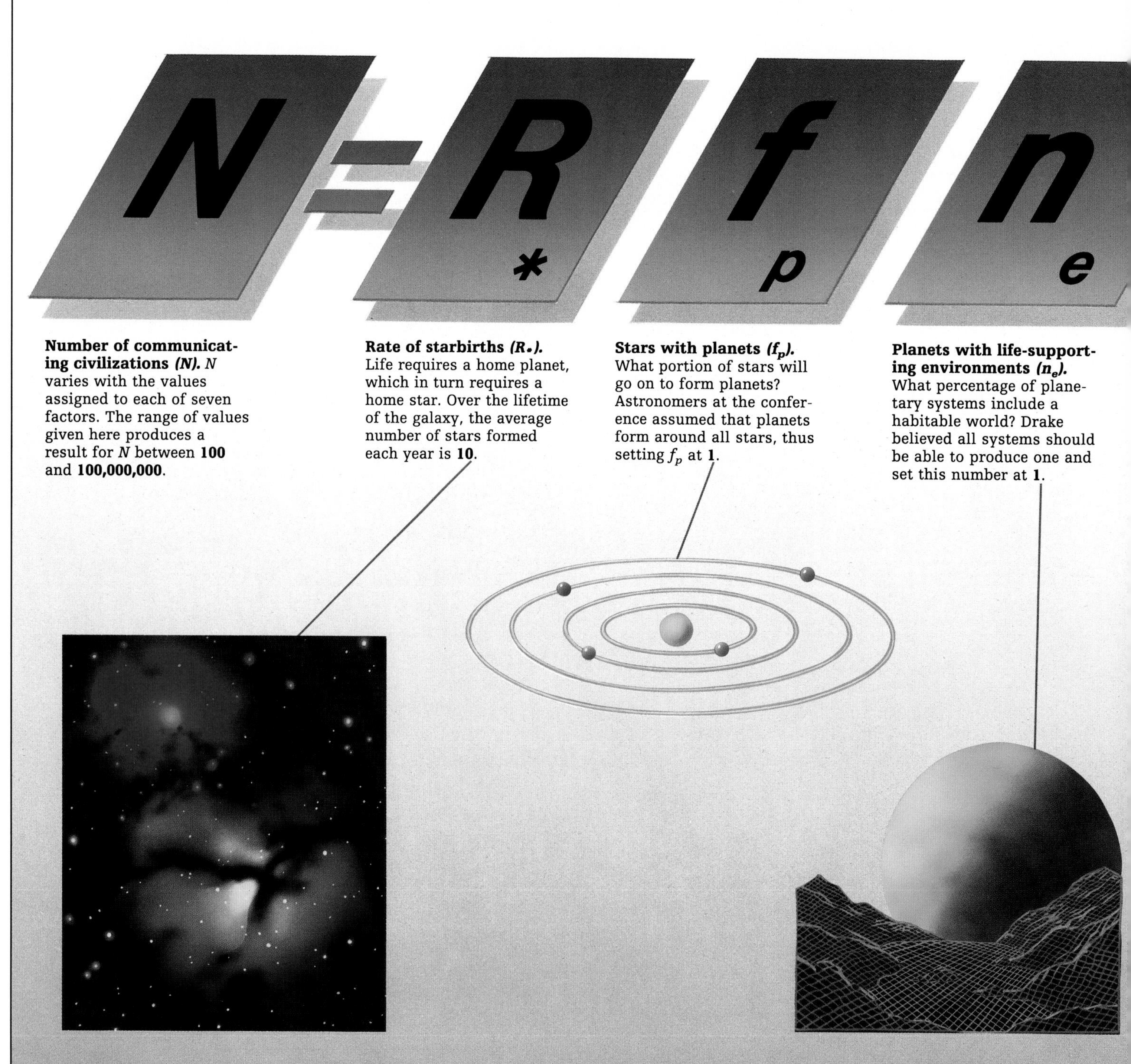

Number of communicating civilizations *(N).* *N* varies with the values assigned to each of seven factors. The range of values given here produces a result for *N* between **100** and **100,000,000**.

Rate of starbirths *(R∗).* Life requires a home planet, which in turn requires a home star. Over the lifetime of the galaxy, the average number of stars formed each year is **10**.

Stars with planets *(f_p).* What portion of stars will go on to form planets? Astronomers at the conference assumed that planets form around all stars, thus setting f_p at **1**.

Planets with life-supporting environments *(n_e).* What percentage of planetary systems include a habitable world? Drake believed all systems should be able to produce one and set this number at **1**.

The first factor in the Drake equation, R_*, is the average rate of star births over the lifetime of the galaxy. Then come five fractional factors—written, by convention, with lettered subscripts—that whittle the total down to the number of stars that may harbor communicating cultures. These are multiplied in turn by *L*, the number of years such a civilization will last after it masters radio communication. Each factor is quite uncertain, making *N* highly speculative. The values below were worked out at a 1961 conference on the possibility of extraterrestrial life, where the equation was first presented; they yield a number that may range anywhere from 100 to 100 million civilizations. Estimates made since then have been similarly varied. Some astronomers believe there is only one communicative world: Earth itself.

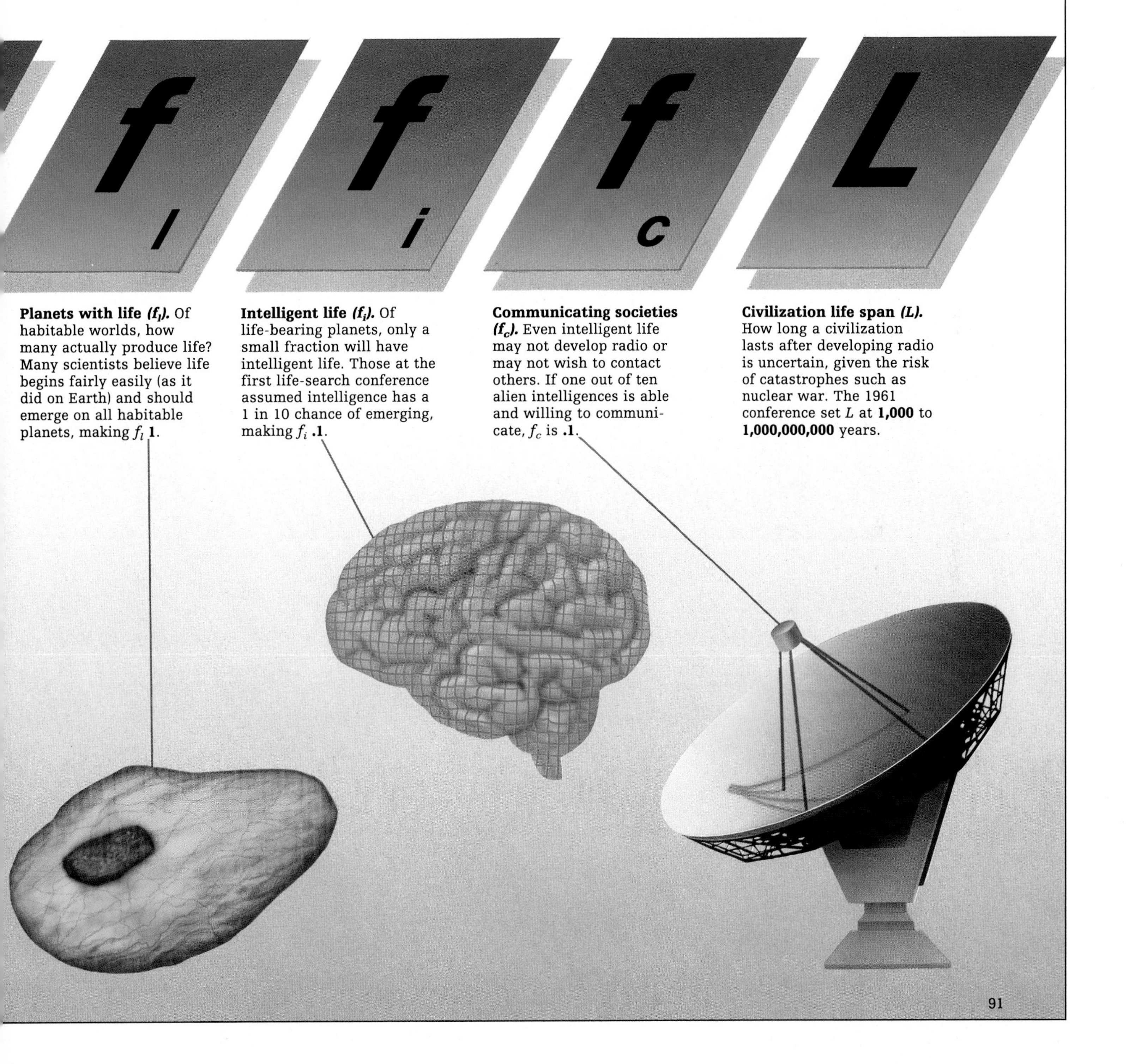

Planets with life ***(f_l)***. Of habitable worlds, how many actually produce life? Many scientists believe life begins fairly easily (as it did on Earth) and should emerge on all habitable planets, making f_l **1**.

Intelligent life ***(f_i)***. Of life-bearing planets, only a small fraction will have intelligent life. Those at the first life-search conference assumed intelligence has a 1 in 10 chance of emerging, making f_i **.1**.

Communicating societies ***(f_c)***. Even intelligent life may not develop radio or may not wish to contact others. If one out of ten alien intelligences is able and willing to communicate, f_c is **.1**.

Civilization life span ***(L)***. How long a civilization lasts after developing radio is uncertain, given the risk of catastrophes such as nuclear war. The 1961 conference set *L* at **1,000** to **1,000,000,000** years.

L

A Galactic Archipelago

No matter how many intelligent species might populate the Milky Way, the likelihood of their conversing with one another depends heavily on the *L*, or longevity, factor in the Drake equation. The reason for this constraint lies in the distances involved in communi-

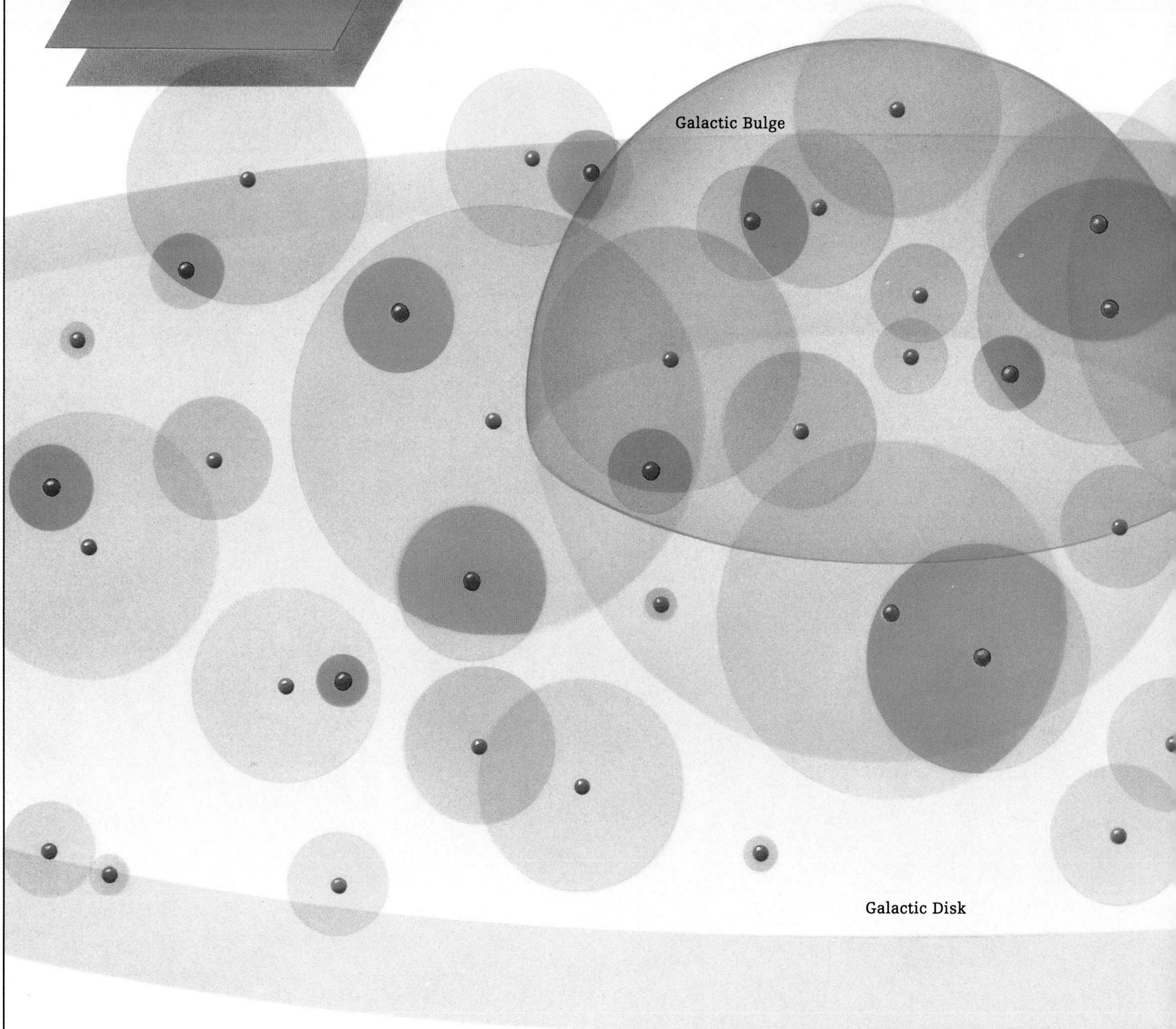

Galactic Bulge

Galactic Disk

cating on a galactic scale. The Milky Way, an average-size galaxy, is 100,000 light-years in diameter. A radio message—which can travel at the speed of light, or 186,000 miles per second—would take 100,000 years to cross from rim to rim. The Drake equation suggests that even inside that span, civilizations are separated by at least hundreds of light-years: Centuries would pass between the sending and receiving of messages.

As for interstellar travel, it would involve immense difficulties even if vehicles could move at light-speed. Thus the galaxy may consist of isolated island worlds *(below)*, each surrounded by the volume of space its radio signals can have reached so far. For example, since Earth discovered radio astronomy relatively recently, its communication zone is unlikely to have extended yet to another inhabited world. Over time, however, as radio signals continue to travel outward, the zone expands—and with it the chance of contact.

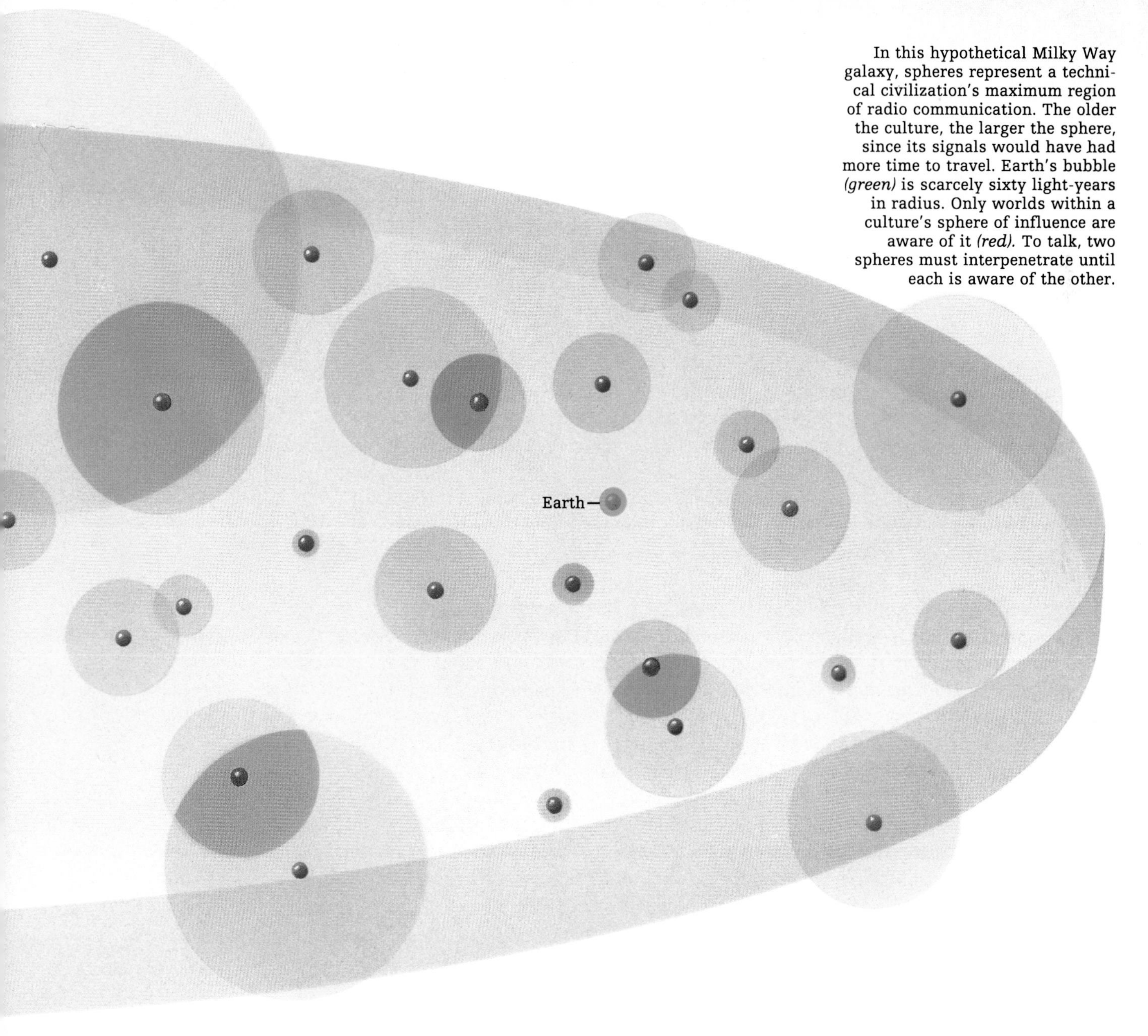

In this hypothetical Milky Way galaxy, spheres represent a technical civilization's maximum region of radio communication. The older the culture, the larger the sphere, since its signals would have had more time to travel. Earth's bubble *(green)* is scarcely sixty light-years in radius. Only worlds within a culture's sphere of influence are aware of it *(red)*. To talk, two spheres must interpenetrate until each is aware of the other.

of organic compounds produced by the interaction of ultraviolet radiation and methane, ammonia, and water. In a tableau described by Dale Cruikshank, Robert Brown, and Roger Clark, methane icebergs float in seas of supercold liquid nitrogen between water-ice continents covered with methane snow. Like Titan and Europa, Triton may be a natural laboratory, richly stocked with the essential ingredients of life.

BARREN WORLDS?

Although such moons are formidable habitats by terrestrial standards, they seem downright welcoming compared to most of the Solar System's planets. Mercury, closest planet to the Sun, is lethally hot by day and lethally cold by night. It is also utterly dry and, with only a trace of atmosphere, a planetary twin of Earth's Moon. Venus, once envisioned as a misty, fecund garden of life, is now known to suffer furnacelike temperatures, crushing pressures, and sulfuric acid rain *(page 89)*.

Mars still retains some of its age-old luster as an abode of life. Huge channels reminiscent of terrestrial riverbeds argue that liquid water was present in significant quantities at some time in the planet's past. Half a decade after Viking sent back its last report from the Red Planet, former Viking biology team leader Harold Klein offered a possible explanation for the space probe's generally unencouraging results: "Chemical evolution on Mars may well have produced living organisms under earlier, more benevolent conditions," he told the NASA historian, "but these organisms couldn't adapt to worsening conditions over geologic time and became extinct." According to Klein and others, Mars may very well be a dead planet today, but one on which a fossil record and organic remains could provide a mute history of past Martian life.

To many researchers, however, particularly in the Soviet Union, Viking's biological results are almost irrelevant. "The fact that nothing has been found at the landing sites of the two American Viking spacecraft is not proof of the absence of life," Valeri Barsukov, director of Moscow's Vernadsky Institute of Geochemistry and Analytical Chemistry, said in 1987. "In the first place, the sites were randomly chosen spots. Besides, one would least expect to find life on the surface. And in the second place, the search technique itself was not very thorough. Therefore, it cannot be ruled out that even on Mars we may find some kind of primitive forms of life."

For that reason, Barsukov added, the Soviet program to investigate Mars in the 1990s includes plans to bring back to Earth material cored out of the Martian permafrost by robots. The Soviet Union began these explorations in July 1988 with the launch of two probes, *Phobos I* and *II,* intended to study Mars and Phobos, one of its small moons. Unluckily, a computer-operator error in September sent *Phobos I* tumbling out of control. *Phobos II* reached Mars, but on March 27, 1989, as the probe closed with its target moon, its transmissions mysteriously ceased, ending the mission.

Beyond Mars, the asteroids and far planets appear to offer only organic chemistry and play for the imagination. Jupiter, larger than all the other planets in the Solar System put together, is a massive ball of liquid hydrogen, with no solid surface to speak of. Nonetheless, as early as 1959, the respected Soviet astronomer Iosef Shklovskii proposed whole Jovian ecosystems, a notion that was later developed by Cornell University's Carl Sagan, along with colleague Edwin E. Salpeter.

According to Sagan, molecules synthesized in the atmosphere might float, planktonlike, at higher altitudes, where temperatures and pressures are less destructive. If such hypothetical proto-organisms could move about and coalesce, they might grow into larger entities. Sagan envisioned organisms in the form of thin, gas-filled balloons, "sinkers," "floaters," and "hunters" occupying special niches in a bizarre Jovian ecology. Such atmospheric creatures might also evolve on Saturn, the other gas-giant planet in the Solar System. But the smaller giants, Uranus and Neptune, seem to offer no possibilities beyond those of rudimentary organic chemistry. Distant Pluto, too far from Earth for close study, likewise appears to offer no animating spark for its cargo of organic chemicals.

On the face of it, the Solar System, except for the haven of Earth, would appear to be inimical to carbon-based life as we know it. But scientists do not rule out the possibility that life forms based on some other chemical—and adapted to wildly different temperatures and pressures—might evolve elsewhere. The life-bearing potential of other members of the Sun's family remains to be seen.

Meanwhile, earthbound scientists continue to scour the infrared sky for other planetary families. A tool that will greatly aid this quest is the *Space Infrared Telescope Facility,* or *SIRTF,* which NASA plans to launch at the end of the century. Sharper eyed, many times more sensitive, and longer lived than the *Infrared Astronomical Satellite,* this Earth-orbiting observatory offers hope that those other Earths, if they are there, will at last be flushed out of the light from their Suns.

A RANGE OF HABITATS

rom the methane frosts of Pluto to the heat-seared craters of Mercury, the Sun's planets and their moons present a diverse collection of habitats. Although Earth is the only known life-bearing body, any world in the solar family might offer interesting possibilities to a visitor from another sun.

The destination of choice would depend on the traveler. To a cool-blooded nitrogen creature, for instance, Triton's brisk minus-360-degree-Fahrenheit surface temperature might seem especially congenial. For a floating entity, the dense, hydrogen-rich atmospheres of Jupiter, Saturn, Uranus, and Neptune could make up an intriguing four-stop package tour. With these and other imagined voyagers in mind, the following pages provide a kind of scouting report on the Solar System. Employing Earth-based measurement units for convenience, the guidebook's entries summarize the properties of planets and moons that would be of particular note to a roving life form.

Any being planning an extended stay would need to know about a world's natural energy supply—the power source for life's activity and long-term survival. The entity would also require water or some other liquid medium to mediate its own complex chemical reactions. And, although life can be astonishingly adaptable, the organism would probably prefer to avoid the stress of a highly inconstant environment. Technologically adroit creatures might share a fourth need: solid surfaces to support whatever they want to build.

The desired temperature, pressure, gravity, and atmosphere would depend on what the species has adapted to. Such conditions derive mainly from astronomical properties like those given below. For example, distance from the Sun helps set a world's temperature, which in turn determines whether its various substances are solid, liquid, or gas. Mass decides a body's gravity and thus shapes the density, depth, and composition of its atmosphere.

Pluto
Mass: .002 Earth.
Average distance from Sun:
3,656 million miles.

Neptune
Mass: 17.23 Earths.
Average distance from Sun:
2,794 million miles.

Triton
(Moon of Neptune).
Mass: .006 Earth.
Distance same as Neptune.

Mercury
Mass: .056 Earth.
Average distance from Sun:
36 million miles.

Venus
Mass: .815 Earth.
Average distance from Sun:
67 million miles.

Earth
Mass: 1 Earth.
Average distance from Sun:
93 million miles.

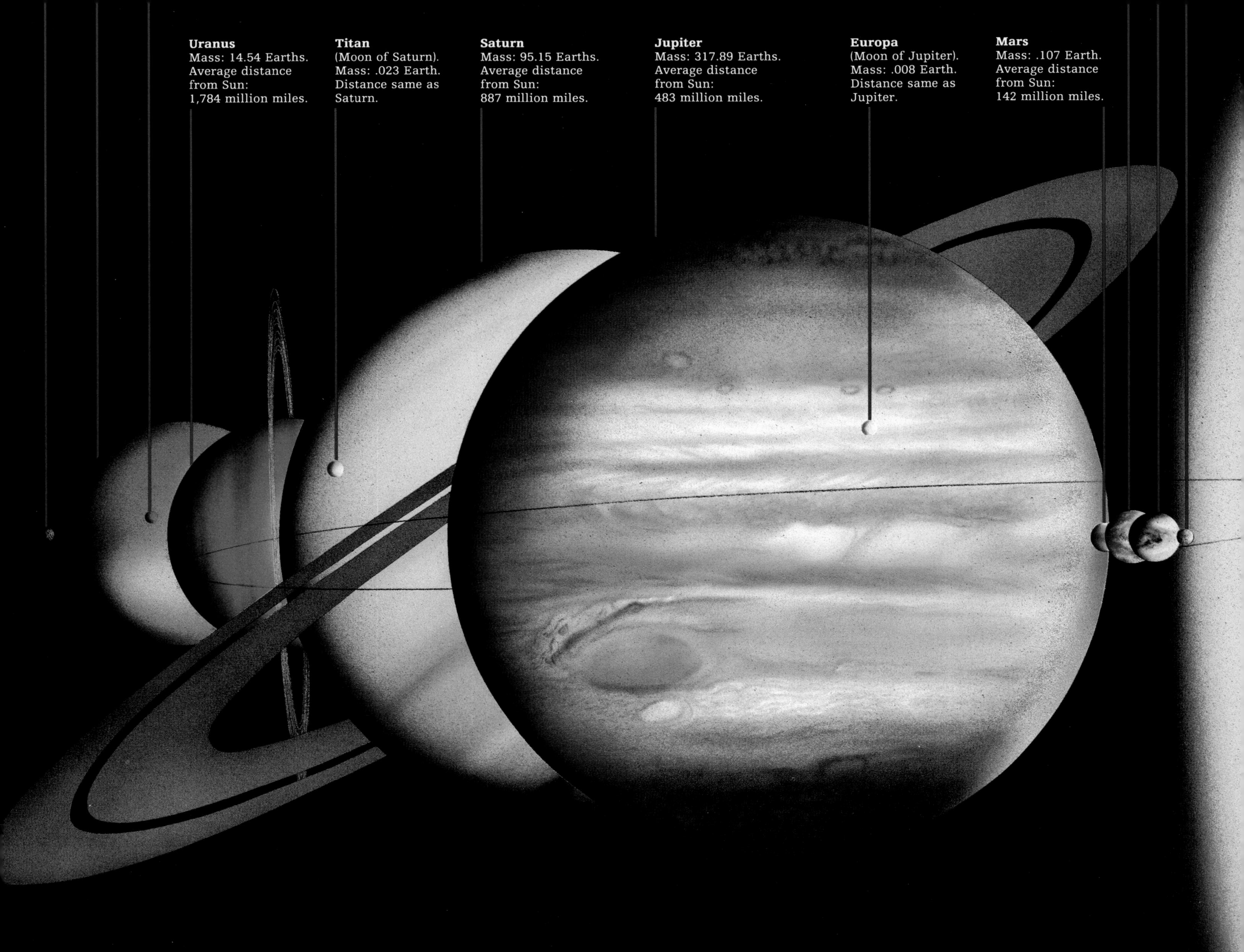
Uranus
Mass: 14.54 Earths.
Average distance
from Sun:
1,784 million miles.
Titan
(Moon of Saturn).
Mass: .023 Earth.
Distance same as
Saturn.
Saturn
Mass: 95.15 Earths.
Average distance
from Sun:
887 million miles.
Jupiter
Mass: 317.89 Earths.
Average distance
from Sun:
483 million miles.
Europa
(Moon of Jupiter).
Mass: .008 Earth.
Distance same as
Jupiter.
Mars
Mass: .107 Earth.
Average distance
from Sun:
142 million miles.

PLUTO

Classification: Small, icy planet.

Atmosphere: Methane, with traces of free hydrogen, perhaps other carbon compounds.

Liquid medium: No surface liquids. Underground nitrogen or hydrocarbon reservoirs possible.

Surface: Relatively smooth water ice, methane ice, possible carbon compound deposits.

Energy: Faint sunlight, averaging .078 watt per square foot, or less than one-thousandth the solar radiation received on Earth. No other known natural energy sources.

Temperature: Between −375 and −350 degrees Fahrenheit, below the freezing point of liquid nitrogen but above the freezing point of liquid hydrogen.

Atmospheric pressure: Less than .01 Earth atmosphere.

Length of day: 6.39 Earth days.

Seasons: Seasonal range of 25 degrees, with each of four seasons lasting 61.75 Earth years. During winter, atmospheric methane frost over existing surface ice, potentially evaporating in summer.

Remarks: Despite possible presence of complex hydrocarbons, indigenous life is highly unlikely, given the limited natural energy sources and low temperature. One to three billion years ago, tidal interactions with the moon Charon may have heated Pluto's interior to produce underground water-ammonia oases. Lasting several million years, these may have provided sites for the development of past life.

NEPTUNE

Classification: Fluid planet, subgiant class.

Atmosphere: Mixture of hydrogen, helium, methane, ethane, and possibly ammonia. Marked by constantly changing cloud patterns, strong winds, ice-crystal haze. Shown here in a 120-mile-thick slice starting from arbitrary zero point in atmosphere.

Liquid medium: Atmosphere gradually compressing to liquid form at its base.

Surface: Fluid-covered, with no known solid surface. Possible ice-coated rocky core deep in planet.

Energy: Sunlight. Stored internal heat, causing planet to radiate 2.5 times its received solar energy, maintaining overall temperature similar to inner neighbor Uranus.

Temperature: −360 degrees Fahrenheit in upper atmosphere, heating to several thousand degrees at lower levels.

Atmospheric pressure: Increases with depth to 6 million Earth atmospheres at core.

Length of day: 17.7 hours.

Seasons: Four seasons, each 41 Earth years.

Remarks: Neptune is cold except at great depths, with constant wind shear creating disturbances between the atmospheric zones. An organic haze layer in the atmosphere suggests a primitive prebiological chemistry involving methane and ethane. The planet is unlikely to harbor native life, however.

TRITON

Classification: Moon of Neptune.

Atmosphere: Methane and perhaps nitrogen. Organic polymer haze at upper levels.

Liquid medium: Possible pools or seas of liquid nitrogen, with some methane. Hints of very low temperature liquid hydrocarbons underground.

Surface: Methane ice and perhaps water ice, with traces of organic compounds.

Energy: Sunlight. Tidal heating from the gravitational stress of an eccentric orbit counter to the planet's spin. Possible radiation from high-energy particles trapped in belts in Neptune's outer magnetosphere.

Temperature: Between −380 and −330 degrees Fahrenheit.

Atmospheric pressure: Uncertain; probably between .00001 and .001 Earth atmosphere.

Length of day: 5.9 Earth days.

Seasons: Seasonal variations of up to a few tens of degrees, with each of four major seasons lasting 41 Earth years.

Remarks: Triton offers areas for settlement equivalent to a small planet's, with pools and seas that may be suitable for surface navigation. Extreme cold appears to rule out indigenous life, although methane and nitrogen may form complex hydrocarbon chains and other biochemicals.

120

105

Ultraviolet Aurora

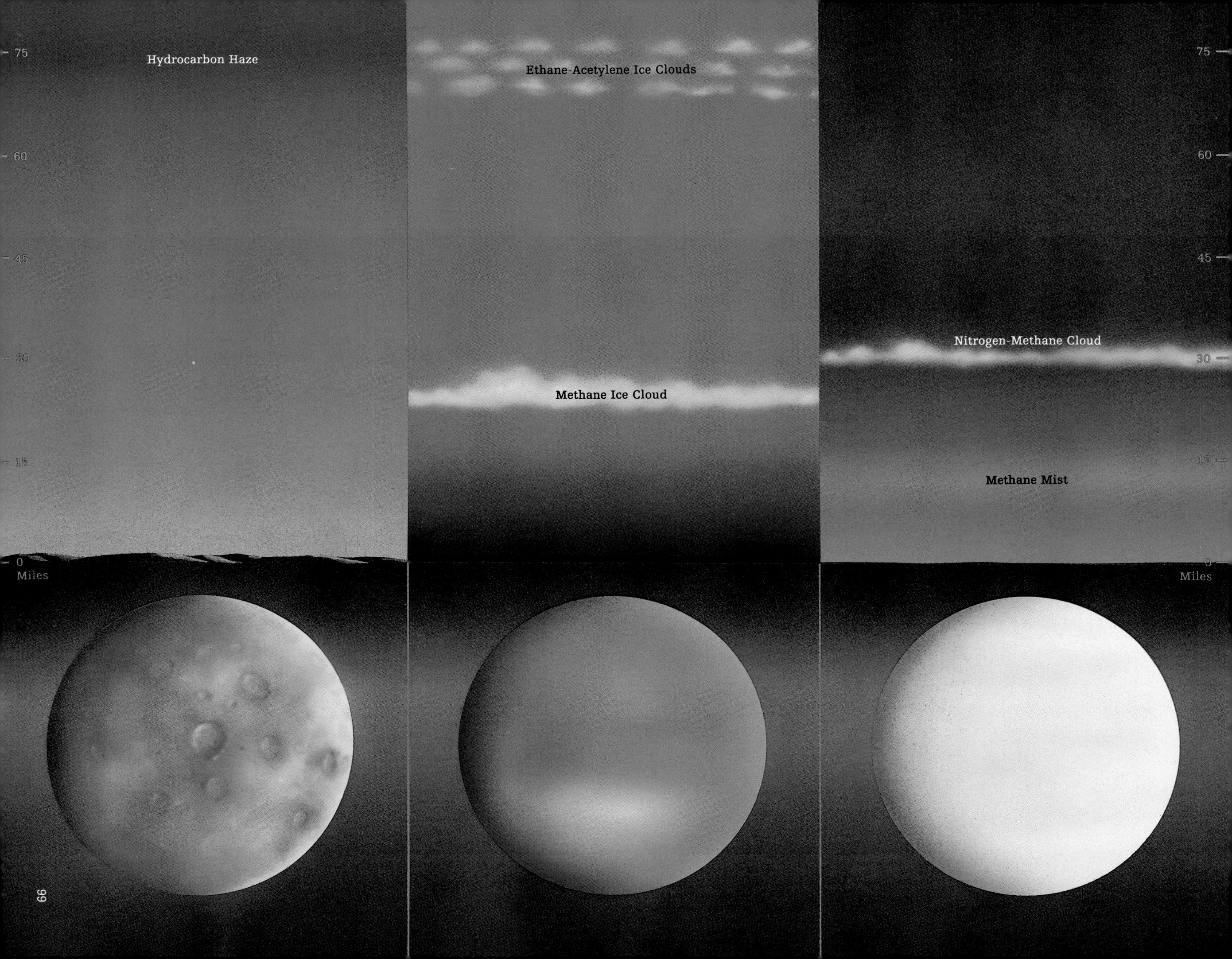

Hydrocarbon Haze
Ethane-Acetylene Ice Clouds
Methane Ice Cloud
Nitrogen-Methane Cloud
Methane Mist
75
60
45
30
15
0
Miles

URANUS

Classification: Fluid planet, subgiant class.

Atmosphere: Hydrogen and helium, with traces of methane, acetylene, and other hydrocarbons, and 90- to 360-mile-an-hour winds. Zero point arbitrary.

Liquid medium: Atmosphere gradually compressing at base to liquid form.

Surface: No solid surface known. Possible rocky core under planetwide liquid cover.

Energy: Sunlight (approximately .0025 the solar radiation received on Earth). Ultraviolet "electroglow" from charged particle clouds in upper atmosphere. In contrast to other fluid planets, minimal heat radiation from core.

Temperature: Temperature inversion in outer atmosphere. Temperatures of −260 degrees Fahrenheit in uppermost atmosphere, falling at lower levels to −360 degrees, then rising again to several thousand degrees near liquid layer.

Atmospheric pressure: Increases to 200,000 Earth atmospheres at liquid layer.

Length of day: 17 hours, 14 minutes.

Seasons: Sunlight and darkness "seasons" at poles, each lasting 42 Earth years, because of planet's axis tilting 98 degrees from its orbital plane. Two summers and two winters per Uranian year at equator. Little seasonal temperature change.

Remarks: Despite probable prebiological organic chemicals, extreme cold in outer layers makes carbon-water life unlikely.

Ultraviolet Aurora

Hydrocarbon Haze

Ethane-Acetylene Ice Cloud

SATURN

Classification: Fluid planet, giant class, with extensive ring system.

Atmosphere: Hydrogen and helium; traces of ammonia, methane, ethane, ethylene, acetylene, and phosphine. Large-scale clouds and storms, equatorial easterly winds of up to 1,100 miles an hour. Hydrocarbon haze at high altitude. Zero point arbitrary.

Liquid medium: Possible water droplets at high altitudes. Compressed liquid hydrogen at great depth, too dense to serve as medium for chemical reactions.

Surface: No solid surface known.

Energy: Sunlight. Heat stored at core from Saturn's formation and continued gravitational contraction, causing planet to radiate about 1.8 times the energy received from the Sun.

Temperature: Ranges from −310 degrees Fahrenheit in upper atmosphere to several thousand degrees in liquid layer.

Atmospheric pressure: From .5 to 1 Earth atmosphere at upper cloud deck to 3 million Earth atmospheres at liquid base.

Length of day: 10 hours, 39 minutes.

Seasons: Four seasons, each about 8 Earth years; seasonal effects strong in upper atmosphere, minimal in lower atmosphere. Wind shear between atmospheric zones, creating turbulent mixing.

Remarks: Life is thought unlikely, but clouds may include prebiological hydrocarbons.

Ultraviolet Aurora

Hydrocarbon Haze

Ammonia Ice Clouds

TITAN

Classification: Moon of Saturn.

Atmosphere: Thick. Nitrogen and methane, with significant hydrogen and carbon monoxide, traces of carbon dioxide and hydrogen cyanide, and such hydrocarbons as ethane, propane, and ethylene.

Liquid medium: Methane rain and probable ocean of ethane, methane, and nitrogen, rich in organic compounds. Possible layer of ammonia and water underlying floor of ocean.

Surface: Ocean or lakes, with exposed water-ice crusts coated with hydrocarbons that have rained out of atmosphere.

Energy: Sunlight. Radiation from particles trapped in Saturn's outer magnetosphere. Heat from tidal flexing. Hydrocarbon fuel in natural gas liquefied on surface.

Temperature: −290 degrees Fahrenheit near equator, −305 degrees at the poles.

Atmospheric pressure: 1.5 Earth atmospheres.

Length of day: Uncertain; estimated at 15 Earth days, 23 hours.

Seasons: Thick atmosphere buffers surface conditions from seasonal change.

Remarks: Titan is similar to Earth in its abundance of biochemical compounds; these may include sugars and amino acid chains. Indigenous carbon-based life is unlikely because of low temperatures. The moon is an ideal natural lab for observing prebiological chemical processes.

75
60
45
30
15
0
Miles
Methane Ice Cloud
Cloud Layer
Ammonium Hydrosulfide
Ice Clouds
Ammonia-Water Ice Clouds
Organic Condensate Haze
Methane Cloud

JUPITER

Classification: Fluid planet, giant class.

Atmosphere: Nine-tenths hydrogen, one-tenth helium, with traces of methane, ammonia, phosphine, water, and hydrocarbons. Large-scale clouds, eddies, and storms. Powerful heat-transport currents circulating between core and outer atmosphere. Zero point here at arbitrarily chosen level in atmosphere.

Liquid medium: Possible water droplets in upper atmosphere. Dense liquid hydrogen at great depth.

Surface: Liquid hydrogen at base of atmosphere, possibly covering solid core.

Energy: Sunlight. Heat stored in core from Jupiter's formation and continued contraction, causing planet to radiate 1.7 times the energy it receives from the Sun.

Temperature: Generally increasing with depth from a low of −280 degrees Fahrenheit to several thousand degrees at liquid hydrogen layer. Strong temperature inversion at upper levels, with gases as warm as −190 degrees.

Atmospheric pressure: Varies from half an Earth atmosphere in the outer clouds to 3 million atmospheres at liquid level.

Length of day: Shortest day of this solar system; 9 hours, 50 minutes.

Seasons: Minimal seasonal change.

Remarks: Because hydrocarbons and organic aerosols exist in the upper atmosphere, indigenous microorganisms are possible, if capable of constant gaseous suspension. However, strong vertical currents pose a threat to any organisms' survival.

— 120

Ultraviolet Aurora

— 105

— 90

EUROPA

Classification: Moon of Jupiter.

Atmosphere: None.

Liquid medium: Possible water ocean underneath icy surface, created by heat generated from tidal flexing. Ocean perhaps as much as 30 miles deep.

Surface: Water-ice sheet, 3 to 50 miles in thickness, marked by extensive fracturing. Relative absence of craters, suggesting recent remelting of icy surface.

Energy: Sunlight. Heat produced by tidal flexing. Possible seafloor volcanism. Charged particles from Jupiter's radiation belts, raining on ice but not affecting subice ocean.

Temperature: Range on ice surface from −340 degrees Fahrenheit at poles to −265 degrees near equator, with noon-to-midnight variation of 90 degrees.

Atmospheric pressure: None.

Length of day: 3 Earth days, 13 hours, 12 minutes.

Seasons: None.

Remarks: If present, the subice water ocean is one of two such environments in the Solar System, the other being found on Earth. Such an ocean may provide a suitable environment for aquatic life, perhaps sustained by heat from ocean-floor volcanism and by sunlight entering through the surface fractures. Indigenous life is highly unlikely on the ice itself.

MARS

Classification: Mid-size rocky planet.

Atmosphere: Carbon dioxide, with traces of nitrogen, argon, oxygen, ozone, and carbon monoxide. Traces of water vapor. Water-ice clouds, low-lying fog at dawn, carbon dioxide clouds. Local high-velocity winds, planetwide dust storms.

Liquid medium: Minimal to none. Possible subterranean deposits of liquid water at great depth. Dry channels suggest catastrophic flooding 3 billion years ago and standing water in craters at earlier time.

Surface: Soil containing bound oxygen, salts, nitrogen, and metallic compounds; probable permafrost under soil. Polar caps of carbon dioxide and water ice. Extensive surface features, including volcanoes, craters, runoff channels, dry riverbeds, volcanic plains, canyons.

Energy: Sunlight with unfiltered ultraviolet rays. Many large, now-dormant volcanoes, suggesting intermittent activity.

Temperature: −190 to 70 degrees Fahrenheit.

Atmospheric pressure: .007 Earth atmosphere.

Length of day: 1 Earth day, 36 minutes.

Seasons: Winter-summer cycle, marked by changes in size of polar caps, seasonal dust storms.

Remarks: Present-day Mars seems unsuited to indigenous carbon-based life. The planet might support microbes in warm subsurface regions where ice would melt to water. Microscopic fossils from more habitable eras may exist. Scattered technical artifacts on the surface originated on Earth.

75
Organic Polymer Haze
Organic Polymer Haze
60
45
Ammonia Ice Clouds
30
Ammonium Hydrosulfide Ice Clouds
Water Ice Clouds
15
Water-Ammonia Fog
0
Miles
75
60
45
Carbon Dioxide Ice Clouds
30
Haze
15
Water Ice Clouds
0
Miles

EARTH

Classification: Large, rocky planet.

Atmosphere: Three-quarters (78 percent) nitrogen, 21 percent oxygen, with traces of other components. Three-atom oxygen, or ozone, blocking short-wavelength ultraviolet radiation from Sun. Marked by long-term wind patterns and extensive water clouds.

Liquid medium: Liquid water oceans as well as rivers and lakes over most landmasses. Widespread, frequent water rain.

Surface: Primarily ocean; one-fourth occupied by soil-covered rock. Extensive water-ice polar caps.

Energy: Sunlight. Lightning. Volcanic activity. Extensive hydrocarbon deposits. Naturally occurring uranium, lithium, deuterium.

Temperature: −125 to 122 degrees Fahrenheit.

Atmospheric pressure: 1 Earth atmosphere at sea level, approximately 14.7 pounds per square inch.

Length of day: 24 hours.

Seasons: Four-season cycle, marked by significant temperature and weather differences.

Remarks: Microbial life has been present for at least 3.5 billion years. The planet has supported abundant plant and animal life, including one technologically adept species approximately 2 million years old. Capacities for interplanetary travel and interstellar communication via radio have recently been developed, but interstellar travel is not yet possible. Approach the planet with caution and tact—no previous contact with other intelligent life is known.

VENUS

Classification: Large, rocky planet.

Atmosphere: Carbon dioxide, with significant nitrogen component and traces of argon, oxygen, carbon monoxide, neon, water vapor, and corrosive acids such as sulfuric acid, hydrochloric acid, and hydrofluoric acid. Thick, permanent cloud cover.

Liquid medium: No surface liquids. Sulfuric acid droplets in clouds.

Surface: Granite, basalt, and possible lava flows, with extensive highland regions and volcanic craters. Extremely diffuse sunlight, providing limited visibility.

Energy: Sunlight. Possible episodic volcanism.

Temperature: At the surface, a fairly constant 900 degrees Fahrenheit produced by runaway greenhouse effect.

Atmospheric pressure: 90 Earth atmospheres.

Length of day: 243 Earth days; 18 days longer than Venusian year.

Seasons: None.

Remarks: From a distance, Venus appears hospitable to carbon-based species, but high pressures, high temperatures, and an extremely corrosive atmosphere probably preclude such life forms. Microbial life, if resistant to sulfuric acid, is possible. Nonindigenous technical artifacts on the surface originated on the third planet, Earth *(left)*, which Venus may have resembled in an earlier epoch.

MERCURY

Classification: Small, rocky planet.

Atmosphere: Extremely sparse atmosphere of atomic oxygen, sodium, helium, hydrogen, and potassium. Occasional light-whitish haze—possibly dust or material released from the interior.

Liquid medium: None.

Surface: Abundant dust over rocky, well-cratered surface; scattered volcanic plains. Half-mile-high cliffs hundreds of miles long, fractures, mountains, ridges.

Energy: Intense sunlight, averaging 871 watts per square foot, about 6.7 times the solar radiation received on Earth.

Temperature: Ranges from −280 degrees Fahrenheit at night to 800 degrees during the Mercurian day—the widest temperature change of any solar planet.

Atmospheric pressure: No more than .0000000000002 Earth atmosphere.

Length of day: 176 Earth days.

Seasons: None.

Remarks: Extreme temperatures and a lack of significant atmosphere or liquid medium make indigenous life unlikely. Little change is expected, since Mercury has been geologically inactive for more than 3 billion Earth years.

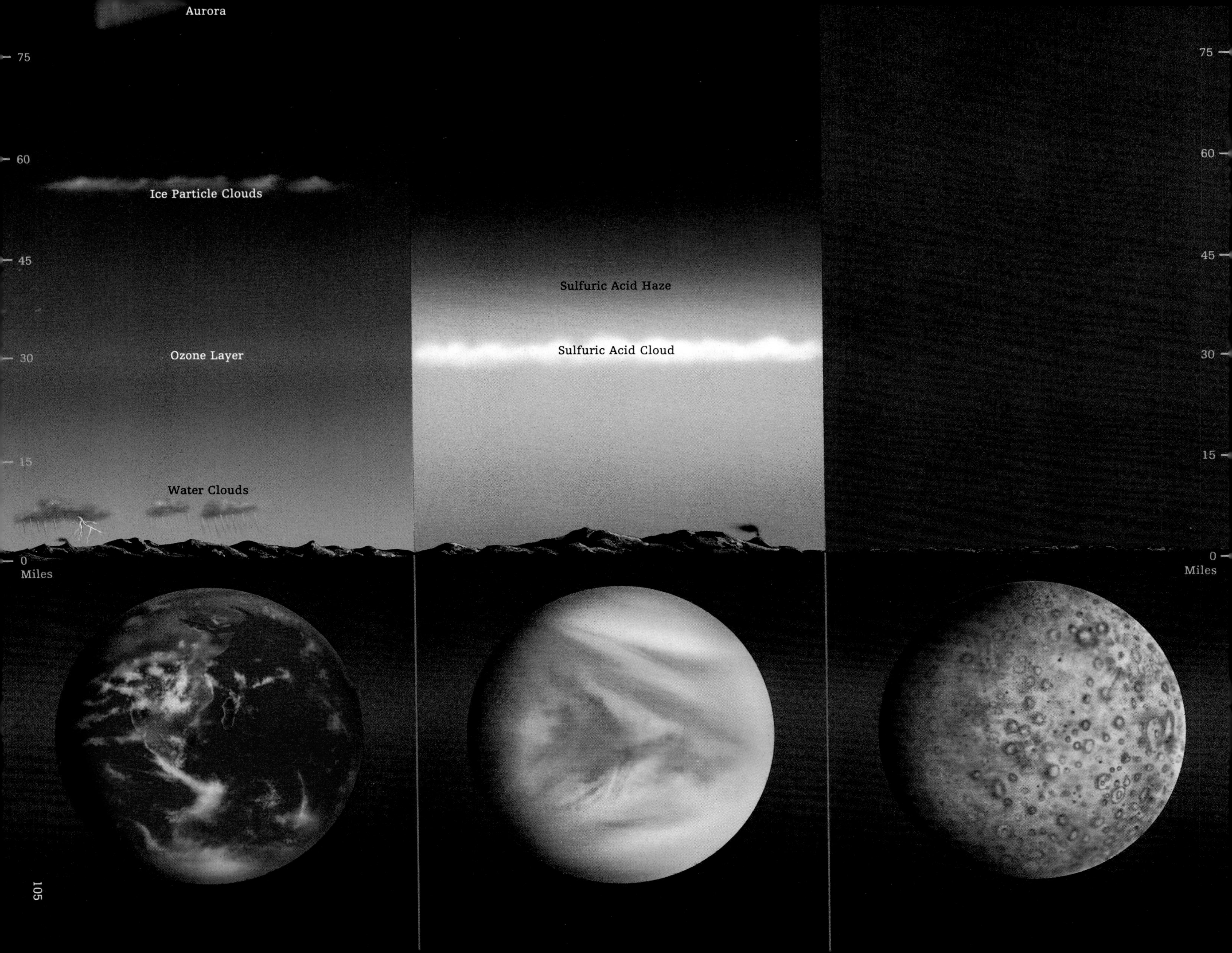

Aurora

75

60

Ice Particle Clouds

45

Sulfuric Acid Haze

Ozone Layer

Sulfuric Acid Cloud

30

15

Water Clouds

0
Miles

75

60

45

30

15

0
Miles

4/Cosmic Companions

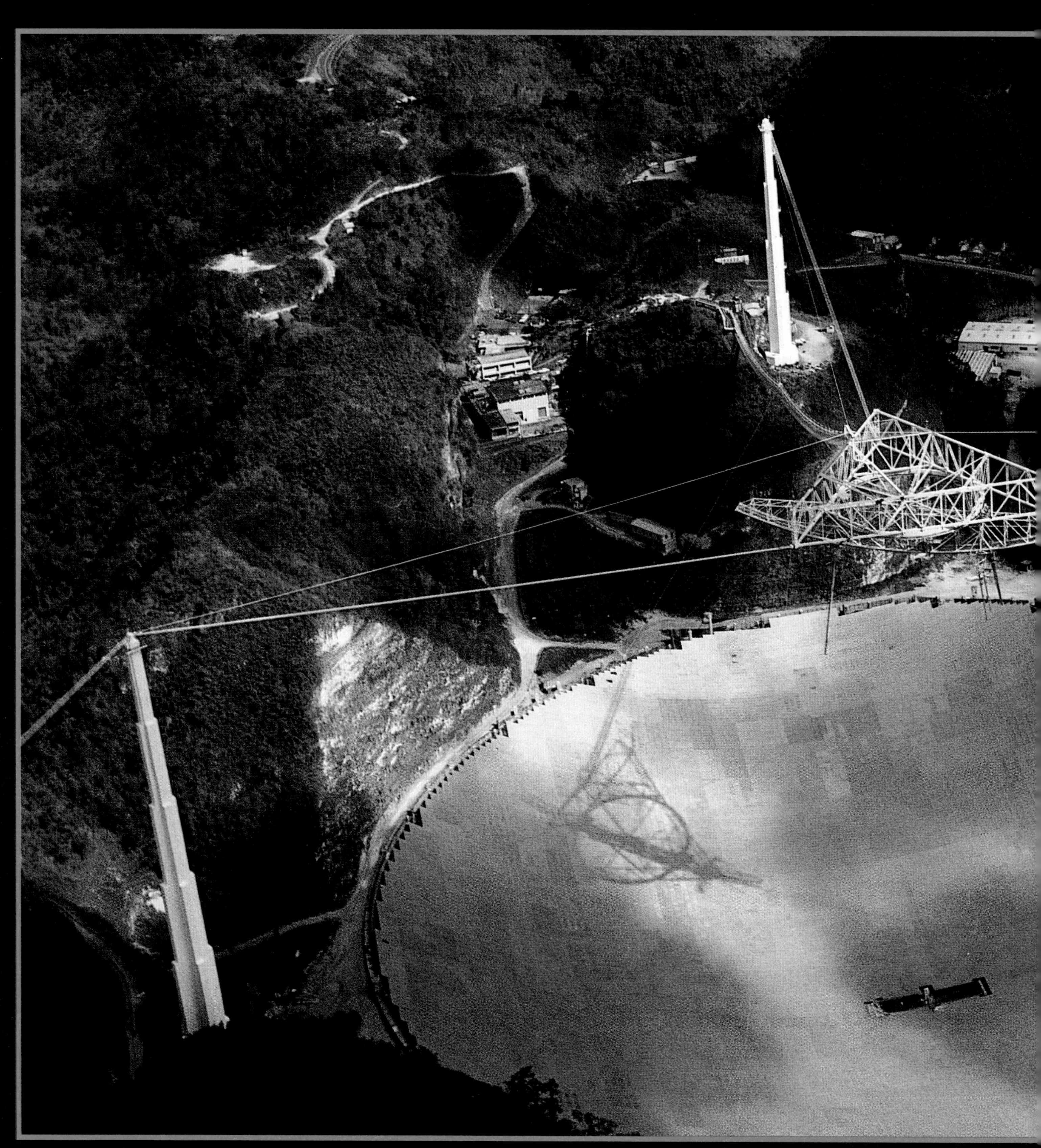

Attuned to the whisper of radio energy from the stars, the National Astronomy and Ionospheric Center's twenty-acre antenna, located near the town of Arecibo, Puerto Rico, is the largest radio telescope in the world. During 1974 astronomers used the instrument to beam a message at a star cluster 24,000 light-years away, in hopes of making contact with extraterrestrial intelligence.

Two nineteenth-century proposals—both probably apocryphal—addressed the formidable problem of initiating communication with whatever sentient beings might be out there in the universe. It is said that in 1820 Karl Friedrich Gauss, a German mathematician famous for his work on number theory and electromagnetism, recommended that a giant right triangle of pine trees be planted in the Russian wilderness, an exercise that would demonstrate to otherworldly observers that Earth harbored creatures civilized enough to understand the Pythagorean theorem. Twenty years later, according to another legend, Viennese astronomer Joseph von Littrow proposed a similarly large-scale stunt: digging a twenty-mile-long ditch in the Sahara, filling it with a tiny ocean of kerosene, and setting it aflame at night.

These extravagant plans took for granted that the mental functions of any alien beings would be sufficiently like ours that they would recognize the signals as a form of communication. But any such assumption about extraterrestrial intelligence may be nothing more than earthly hubris. Terrestrial scientists are just beginning to explore the real nature of intelligence as it exists on this planet, and they have no clear conception of how minds might be shaped by other worlds. Nevertheless, they listen hopefully for evidence of thinking creatures amid the various radio whispers and whistles emanating from space. No one knows whether such contact would be friendly or ultimately fearsome, a greeting from a compatible culture or a kind of trumpet blast from technologically powerful aliens to whom human society might seem aboriginal and immature. If threat exists, there is no hiding from it: Earth's signal has already been sent.

A BRIGHT BEACON

The reported suggestions by Gauss and von Littrow for turning portions of the planet into a signal beacon were never put into effect, but in the century and a half since, a signal beacon is exactly what Earth has become. Instead of artificially shaped pine forests or mammoth bonfires, however, the signal consists of decades' worth of radio waves.

Once humans began broadcasting at radio frequencies in the early twentieth century, their assorted messages—radio, radar, and then television—began leaking into space, creating a brilliant bubble of detectable energy that is expanding outward at the speed of light *(pages 92-93).* Cornell University astronomer Carl Sagan has pointed out that an extraterrestrial astronomer looking at the Solar System today would see an odd pair of radio sources orbiting one another: one, an unprepossessing yellow dwarf star, radiating a normal amount of energy at the radio end of the electromagnetic spectrum; the other, a body about one three-hundred-thousandth the star's mass but ten to a hundred times brighter at radio wavelengths. By the end of the twentieth century, Earth's radio sphere of influence will be some 100 light-years across. Whether those signals will reach distant shores equipped with instruments capable of receiving them and whether the masters of those instruments will wish to follow up on the indication of another presence in the neighborhood remain to be seen. Meanwhile, radio astronomy has appropriated the other side of the issue—the earthly detection of communication from afar.

THE WATER HOLE

In 1958 the National Radio Astronomy Observatory was established in Green Bank, West Virginia, one of the first facilities dedicated to examining the universe at the longer wavelengths of the radio region of the electromagnetic spectrum. When the facility's new eighty-five-foot radio telescope was ready a year later, two young astronomers, Frank Drake and David Heeschen, were charged with figuring out how to use it to maximum effect. As Drake remembered thirty years later, the astronomers were eager to test its limits; in such an infant field, they figured, "it was easy to make discoveries."

One day, while calculating the instrument's ability to detect faint signals from space, Drake posed a purely hypothetical problem: At what distance would the telescope be able to detect the radio energy emitted by Earth? "It worked out that it could detect the strongest signals leaving earth from a distance of ten or twenty light-years," he recalled in 1985, "which was pretty remarkable." More remarkable, Drake's calculation suggested that the Green Bank astronomers should be able to hear the radio emissions of any extraterrestrial societies within that same range. "We ought to look," he told his colleagues. "For all we know, practically every star in the sky has a civilization that's transmitting." Otto Struve, the Russian-born director of NRAO, concurred and gave the twenty-nine-year-old Drake the go-ahead to launch a search. The assignment was less straightforward than it seemed at first. Radio waves cover millions of

Modeled on an ancient Greek coin, this pewter "Order of the Dolphin" commemorates a 1961 meeting in Green Bank, West Virginia, where a small group of scholars discussed the likelihood of contact with alien civilizations elsewhere in the galaxy.

frequencies in the electromagnetic spectrum, and Drake was instantly confronted with a troubling question: On what wavelength would alien civilizations broadcast? As the astronomer remembered it years later, the wavelength finally selected was chosen for pragmatic as well as scientific reasons. "We had to be very careful not to do things that would bring criticism," he recalled. "And so, very early on we made two decisions. One was that we would seek no publicity. We would keep the project completely secret as much as possible." The second decision was that their detection apparatus also had to be useful for conventional astronomy. That took them to the twenty-one-centimeter wavelength of neutral hydrogen, a strong feature in radio spectra. There, they reasoned, they would be able to study a recently discovered splitting of the twenty-one-centimeter absorption line caused by the radio energy's interaction with magnetic fields and search for alien broadcasts at the same time.

Some six months after the Green Bank team began building a receiver tuned to a narrow band around twenty-one centimeters, two physicists at the Massachusetts Institute of Technology also proposed a search for extraterrestrial signals just off the same wavelength, but for very different reasons. In a paper published in the scientific journal *Nature,* Giuseppe Cocconi and Philip Morrison urged the choice of that wavelength because there is relatively little background interference, either natural or human-made, in this region of the radio spectrum; any signal detected would thus likely be both artificial and not of earthly origin.

Years later, almost lyrical reasons were given for searching at twenty-one centimeters. NASA physicist Bernard Oliver insisted that scientists picked the area around twenty-one centimeters because it falls between the wavelengths of hydrogen (H) and the hydroxyl radical (OH), the ingredients of water (H_2O). "Standing on either side of a gate," Oliver said, the two markers on the electromagnetic spectrum would beckon "all water-based life to search for its kind at the age-old meeting place of all species: the water hole."

At six a.m. on April 11, 1960, Drake and his colleagues at Green Bank officially began the search—which they had named Ozma, after the queen of Oz, the fictional kingdom noted for its wizard. The first targets were two Sun-like stars, Tau Ceti and Epsilon Eridani. At 11.9 and 10.8 light-years' distance, they were near enough to Earth for any signals from their vicinity to be easily detected.

The astronomers tried to tune in Tau Ceti first, but after six hours of observation the star seemed to have nothing to say. Next, the team cranked the eighty-five-foot dish around toward Epsilon Eridani. "In less than a minute we suddenly heard something remarkable from a loudspeaker connected to the system," Drake recalled years afterward. "We heard a very intense, pulsed noise." Beating regularly eight times a minute—a regularity not at all characteristic of naturally occurring radio emissions from celestial bodies—the loud signal sent a chill through the astronomers. "We were all dumbfounded," said Drake. "Could it be this easy? All you need to do is point

Conversations with Other Species

If human and alien ever make contact, each will face an extraordinary communication gap. With nothing but intelligence in common, the two must somehow learn to speak together. In planning for that challenge, radio astronomers who scan the skies for signs of extraterrestrial intelligence are paying close attention to research projects investigating the many ways in which animals on Earth communicate.

A key issue for such investigations is determining whether communication is really happening. Animal researchers particularly emphasize this step because of the embarrassing story of Clever Hans *(below)*, a horse whose uncanny sensitivity to body cues falsely suggested it could answer mathematical questions. When legitimate messages do exist, animal researchers must deduce what they mean. Usually this is done by searching for communication patterns: examining the relation between animal sounds and herd movements, for example.

As indicated below and on the following pages, research projects focused on animal communication have yielded plentiful results—from breaking the dance code of bees *(overleaf)* to teaching a gorilla to speak American Sign Language with the fluency of a human four-year-old. Whether these achievements will ever help humans recognize and interpret an extraterrestrial message remains to be seen. For the moment, radio astronomers can only listen and hope.

1 Taught to communicate in American Sign Language, a human-raised gorilla named Koko also uses a keyboard device that emits English words when she selects and depresses the keys *(right)*. By combining several keystrokes, Koko may answer questions, offer comments, or simply transmit a request: "Want apple, want eat."

2 The horse Clever Hans amazed turn-of-the-century Berlin by rapping out solutions to mathematical and musical queries; some answers were interpreted using an adjacent chart. Hans actually knew nothing of square roots and notes. The horse simply tapped a hoof until the interrogator's pose or expression altered.

3 Closely connected to language and intelligence, social behavior is well developed among cetaceans. In the wild, the mother and calf dolphin shown at right may be accompanied by a helpful dolphin "aunt." Safe in captivity, mothers are less defensive but remain close to their calves—who often copy a trained parent's behavior.

Messages without Words

Earthly evolution has produced many techniques for communication. Sparrows convey territorial boundaries in song *(opposite),* passing on melodies from one generation to the next. Other animals, including bats, shrews, and porpoises, manipulate sound at extremely high pitches, and elephants converse in infrasound—notes too low for humans to hear.

Then there are species like the honeybee, which eschews sound altogether. The bees communicate in a formal dance that consists of two circular segments connected by a brief, waggling run. The length of the run represents the distance to a cache of pollen, water, or building supplies, and the run's direction symbolizes the find's location relative to the Sun. The energy of the dancer indicates the material's degree of importance. Because dances occur in dark hive interiors, bees frequently hug the body of the dancer, receiving the message by touch and smell.

1 Bottle-nosed dolphins approach a three-key control board that sends their messages in whistles to a poolside trainer. A triangle key yields a ball, the key with a notched symbol produces a ring, and the letter H key usually gets the dolphins a petting from the human.

2 Vervet monkeys warn of its predators with different signals. Rapid coughing indicates a dangerous snake. A bark—denoting a leopard or other hunting mammal—drives monkeys to the highest branches. Grunts trigger a dive for the bushes: An eagle is attacking.

3 Humpback whales communicate in lengthy, complex "songs" that gradually change during the whales' winter and spring stay in the tropics. The whales head north in silence for summer and fall feeding *(left),* then resume last year's songs upon returning south.

Basic Phrases

Some animals show a surprising ability to converse with humans. Even when the dialogue is one-way, animals in human-language projects readily parse sentences by agreed-upon rules. For the sea lion and dolphin at far right, "ball, ring, fetch" consists of two nouns and a verb and means to fetch the ring to the ball's location. "Ring, ball, fetch" would mean to move the ball to the ring. In two-way communication, language is more than an experimenter's tool. It allows animal and human to exchange information, tell stories, even become friends. Taught sign language, Koko, a gorilla in California, learned about kittens, then asked her human mentors for one of her own. In another project, a parrot called Alex proved capable of naming shapes, colors, and quantities—and learned to convey irritation with the comment "I'm going away."

1 Koko's fondness for cats has inspired her to make repeated sign-language requests for kittens, which she treats with maternal care. Reveling in the fun of naming, Koko dubbed her first pet All Ball, the second *(right)* Lipstick for its pink nose and mouth.

2 The parrot Alex, seen here beak to nose with his human researcher, can say and understand the English words for seven colors, five shapes, twenty distinct objects, and quantities up to ten. Moreover, he readily relates them—stating, for example, that red and green squares differ in color but share the same shape.

3 Koko particularly enjoys storytelling, never tiring of such favorite tales as "The Three Little Kittens." Here, she emphatically gives the sign for "mad" as part of a running commentary on the naughty kittens and their irate mother.

1 Attentive bees press against a scout, just returned to the hive with news of a find such as food or water. The geometry of the scout's waggle dance conveys the material's location; its desirability is directly proportional to the intensity of the scout's performance.

2 As nesting begins, a white-crowned sparrow bursts into song, proclaiming his territory to all within earshot. Since the songs differ regionally, newcomers as well as youngsters must learn the appropriate dialect from local males.

3 Asian elephants *(left)* emit numerous sounds at frequencies below human hearing. Little is known about the infrasonic calls, discovered in 1984. Some are followed by herd redirections and may indicate danger; others seem to be used by the males to locate females during the short breeding season.

Animal Vocabularies

Researchers use many methods in their efforts to understand animal communication. In some cases, they devise a simplified language, like the whistle codes employed in the dolphin project at far left. More often, they observe animals closely in the wild to fathom their natural languages.

Although animal languages lack the complex grammatical rules and extensive vocabularies that distinguish human language, many species do have what can be considered the equivalent of words. Among the vervet monkeys of East Africa *(opposite),* several different sounds for "predator" are used, each indicating a distinct type of threat, such as a python, a jackal, or an approaching human. Researchers have confirmed these meanings by recording the sounds, then playing them back from a speaker concealed in natural vegetation. The monkeys seek cover as though the indicated predators have appeared.

1 A sea lion named Rocky watches carefully as a human researcher signals "pipe" in the sequence "water wing, pipe, fetch," a request to take a pipe to the water wing. The signaler wears opaque goggles to eliminate eye cues, requiring Rocky to interpret the signals alone.

2 A blindfolded handler signals "pipe" differently when asking a dolphin called Akeakamai to take a hoop to the pipe. Like the sea lion at left, Akeakamai responds to such commands even when the combination of signals is new, indicating real linguistic ability.

to a random star and within one minute you get a signal that puts your receiver into overload?" Minutes later, the strange signal vanished. In ten days it was back, but this time the astronomers had a second receiver to help identify it. To their profound disappointment, they determined that the source was terrestrial, probably a transmitter on a passing military aircraft. Although they continued their observations of Epsilon Eridani and Tau Ceti intermittently until July, they detected no further "intelligent" signals.

Unrewarding as Project Ozma was, it stirred imaginations. Radio astronomers all over the world began to listen for alien signals and had some thrilling moments. In 1964, for example, Nikolai Kardashev, then at the University of Moscow's Sternberg Observatory, urged his fellow Soviet scientists to point their radio telescopes at CTA 102, a radio source later identified as one of the very distant objects called quasars, to confirm possible artificial signals. A year later, Kardashev and two colleagues reported such confirmation. "We have recorded radio oscillations of constant period that repeat themselves every hundred days," they wrote in the journal *Soviet Astronomy.* "The data have been checked by our research team with the greatest care and leave no room for doubt." Unfortunately, the findings could not be duplicated, a fate that befell numerous similar reports from the Soviet search. Most of the alleged discoveries could be explained by other phenomena—a peculiar kind of flare activity on the Sun, for instance, or in one tantalizing case, a transmission from what appeared to be a secret U.S. satellite.

A few years later, a young Cambridge University graduate student named Jocelyn Bell found something that at first resisted all obvious explanations. In 1967 Bell was studying readings of the radio sky in the constellation Vulpecula (the Fox) when she detected a totally unfamiliar pattern in what had seemed to be a natural radio burst. The emissions were extraordinarily rapid and supremely regular, pulsing every 1.33728 seconds. Mostly in jest, Bell and her coworkers dubbed the signals LGM, for little green men, causing the public heart to skip several beats. As it turned out, however, Bell had discovered not a phrase of alien Morse code but the first known example of a neutron star, an astonishingly dense, rapidly spinning body whose powerful radio beam sweeps past Earth like the repeating beam of a distant lighthouse. Now known as pulsars, the objects are the closest that terrestrial listeners have come to detecting the sort of coherent, repeating signal that could denote another intelligence. Still, the hunt goes on.

SEARCHING

After Project Ozma and its international counterparts, the search for extraterrestrial intelligence—SETI, as it is called—assumed other forms, always tuned to the radio wavelengths bounding the water hole. At Ohio State University, the longest-running American search yielded a highly regular radio emission in 1977 that has been known ever since as the WOW! signal, for the notation an excited observer wrote in the margin of the data printout. The radio burst did not repeat itself and has never been explained.

NOBODY IS OUT THERE

Tulane University physicist Frank J. Tipler considers the search for extraterrestrial intelligence a waste of time and money. "Interstellar travel would be simple and cheap for a civilization only slightly in advance of our own," he wrote in 1983. Their autonomous space probes would use materials en route to fuel and rebuild themselves. At 100,000 years per interstellar flight and 1,000 years to construct each new probe, Tipler figured a single probe would take 300 million years to send a descendant to every star system in the galaxy. Allowing six billion years from the formation of a planet to the time its intelligent species begins sending out probes, we should have heard by now from anyone whose star system is more than 6.3 billion years old, the age of about half the stars in the galaxy. "If a civilization approximately at our level had ever existed in the galaxy, their spaceships would already be here," he wrote. "Since they are not here, they do not exist."

But all early efforts were severely limited in their ability to search the cosmos for faint alien signatures embedded in the spectrum of natural radio noise. With the advent of powerful computers, the task has become marginally more manageable. In 1983, for example, Paul Horowitz, a forty-year-old physicist at Harvard, built a receiver that, rather than listening at a single wavelength, could scan the spectrum in 131,000 channels in the vicinity of the hydrogen line at twenty-one centimeters—a frequency of 1.4 billion cycles per second, or gigahertz. Called a "suitcase SETI," Horowitz's multichannel receiver could be used with any radio telescope, anywhere in the world, allowing him to piggyback on telescopes while they were serving other purposes. This became the centerpiece of Project Sentinel, which used an eighty-four-foot radio telescope at the Smithsonian's Oak Ridge Observatory in Massachusetts. In 1985 Steven Spielberg, the film director-producer who added "E.T." to the language, funded Horowitz's further development of his computer to monitor 8.4 million narrow radio channels centered on the hydrogen frequency. By the late 1980s, Horowitz's receiver was the heart of META, the Megachannel Extraterrestrial Assay at the Massachusetts facility.

At one point Horowitz had been targeted to receive NASA funding out of two million dollars budgeted for the agency's extraterrestrial-life activities. But in 1981 these plans won the Golden Fleece Award, bestowed upon presumably

Scanning the Skies for Alien Signals

Scientists face some daunting hurdles in their search for a message from beyond. To capture an alien radio signal, a receiver on this planet would have to be pointed in just the right direction and tuned to a precise frequency. Not only could the signal be coming from any one of millions of stars scattered across the sky, but an alien civilization could choose from among billions of frequencies as well. And if humans managed to guess the correct frequency on which an advanced civilization would communicate, the wavelength could be shifted by the motion of the distant world relative to the Earth.

The earliest message hunts were usually restricted to the region of the electromagnetic spectrum near 1.4 gigahertz, the frequency at which interstellar hydrogen emits radiation. The most common element in the universe, hydrogen radiates in an uncluttered part of the microwave region; an artificial signal would thus stand out. Later, scientists concluded that frequencies between hydrogen and the hydroxyl molecule should also be investigated because the two combine to form water, and other life forms might be water-based too. This area of the spectrum came to be known as the water hole. In the 1990s, the National Aeronautics and Space Administration expects to launch a ten-year search that will scan all frequencies across a broad band of the microwave region. The survey will cover the entire sky but will focus most closely on 773 stars that resemble the Sun.

If another civilization in the Milky Way is signaling Earth at the hydrogen frequency, the message will arrive on a different wavelength because of the Doppler shift: The waves will be compressed or stretched by the relative motion of the two worlds. Three principal motions are involved. The planets each orbit a sun *(small circles)*, the suns move with respect to the local group of stars *(small arrows)*, and the local groups of stars circle the center of the galaxy *(large arrows)*.

The Milky Way is so large that by the time a signal from a distant planet reached Earth, both worlds would have changed positions significantly. For example, if a signal were beamed at Earth from a world 10,000 light-years away, the Earth would travel more than 420 trillion miles around the galaxy before the signal arrived. Any aiming of messages thus involves a great deal of advance planning.

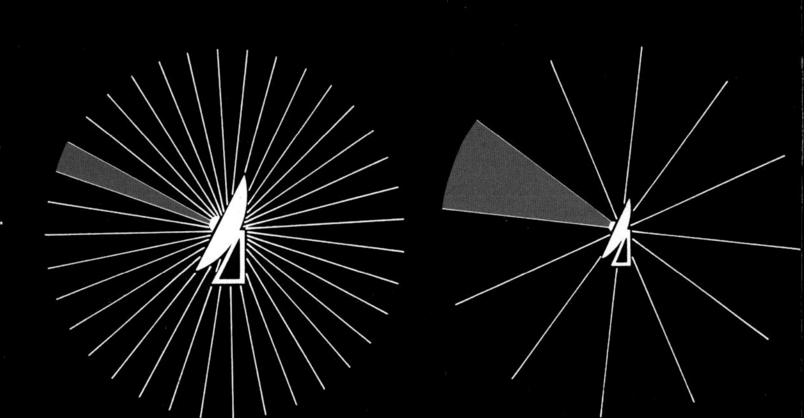

NASA's sky search will use two radio telescopes. One will be a large dish *(near right)*, which is more sensitive to weak signals than a smaller telescope but has a narrower field of view. This will concentrate only on Sun-like stars. The second *(far right)* will be smaller and less powerful but better suited to sweeping the sky because of its large field of view.

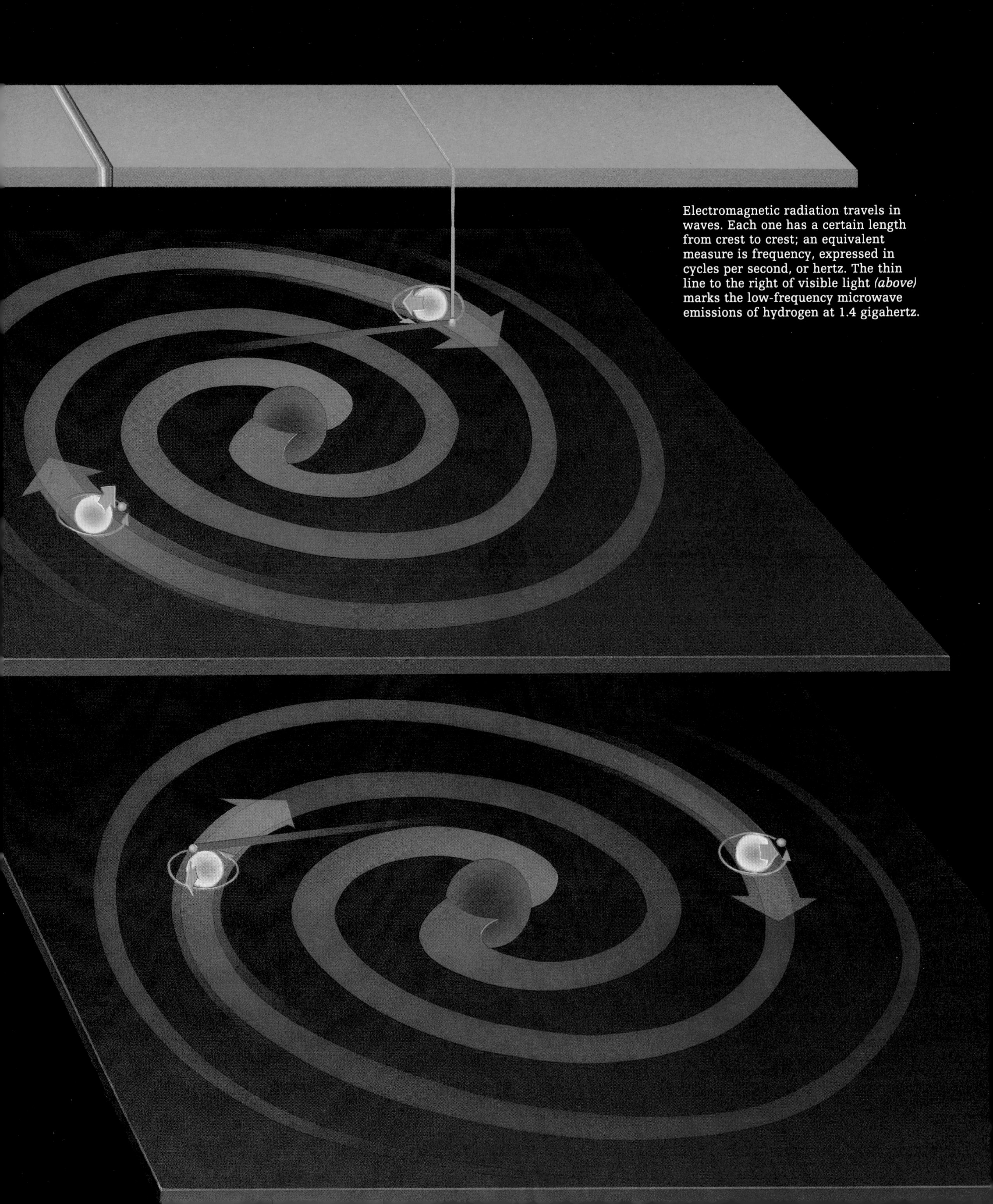

Electromagnetic radiation travels in waves. Each one has a certain length from crest to crest; an equivalent measure is frequency, expressed in cycles per second, or hertz. The thin line to the right of visible light *(above)* marks the low-frequency microwave emissions of hydrogen at 1.4 gigahertz.

The first seekers of radio messages from outer space concentrated on a few likely stars and guessed that the frequency of hydrogen, at 1.4 gigahertz, was a logical place to start. Some searchers broadened the quest to related wavelengths at 1.7 (hydroxyl) and 22 (H_2O) gigahertz. In 1985 astronomer Paul Horowitz of Harvard University began a more sophisticated project he called META (Megachannel Extraterrestrial Array). His equipment scans the entire sky along eight million channels near a chosen central frequency, such as that of hydrogen or hydroxyl. By searching millions of frequencies at the same time, project META has eliminated the problem of the Doppler shift, since its wider range would be able to capture a signal even if it were slightly shifted from its original frequency. But the most comprehensive search of all will come from NASA in the 1990s. The agency's equipment will cover eight million very broad channels and scan the microwave spectrum from 1 to 10 gigahertz.

Early searches for extraterrestrial signals concentrated on three isolated frequencies *(narrow green bands, below):* those of hydrogen, at 1.4 gigahertz; hydroxyl (HO), a molecule composed of one atom of hydrogen and one atom of oxygen, at 1.7 gigahertz; and the water molecule (H_20), at 22 gigahertz.

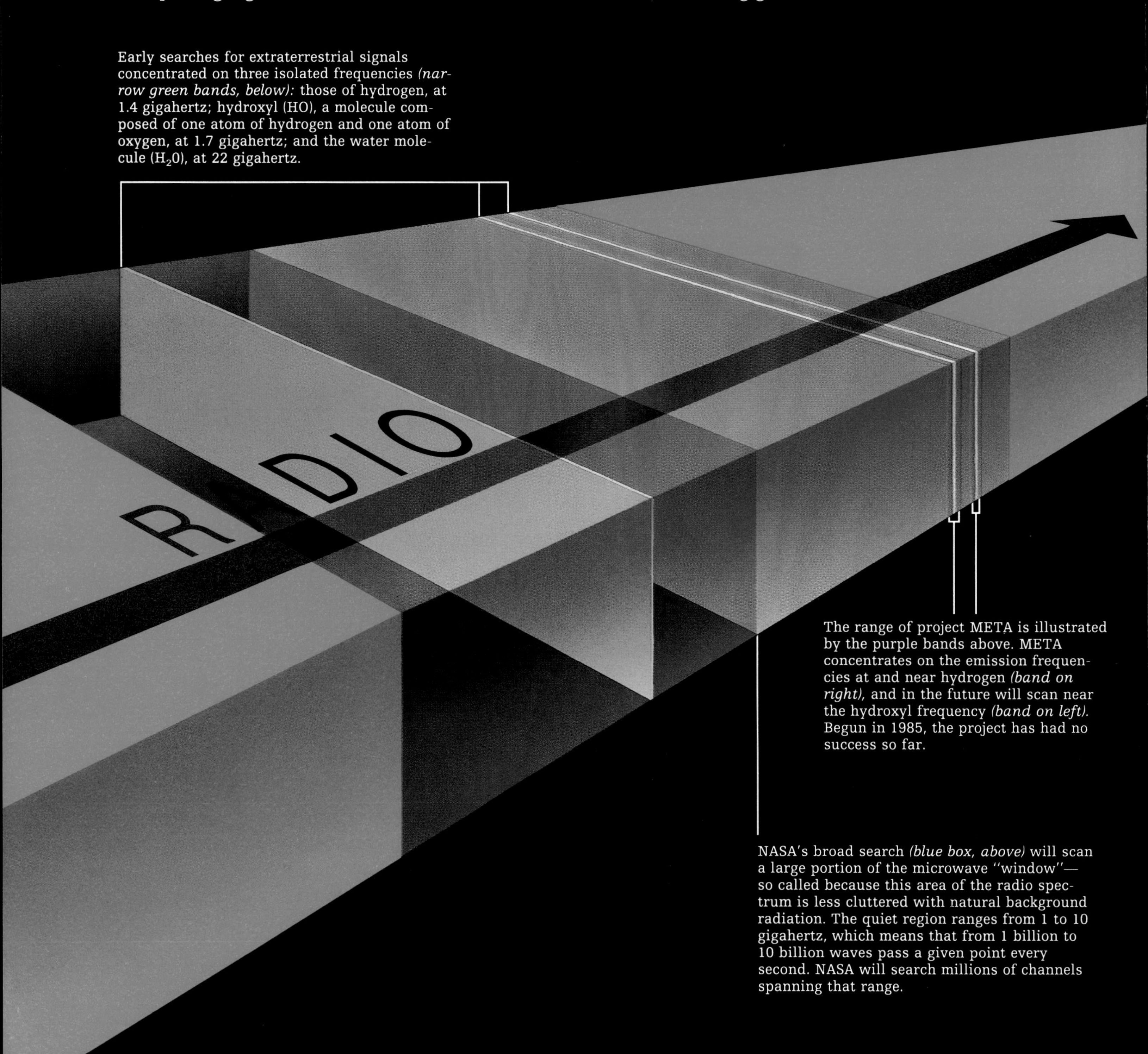

The range of project META is illustrated by the purple bands above. META concentrates on the emission frequencies at and near hydrogen *(band on right),* and in the future will scan near the hydroxyl frequency *(band on left).* Begun in 1985, the project has had no success so far.

NASA's broad search *(blue box, above)* will scan a large portion of the microwave "window"—so called because this area of the radio spectrum is less cluttered with natural background radiation. The quiet region ranges from 1 to 10 gigahertz, which means that from 1 billion to 10 billion waves pass a given point every second. NASA will search millions of channels spanning that range.

wasteful government projects by Senator William Proxmire of Wisconsin. He also amended the agency's 1981 budget to prohibit all such work. Funding was restored in 1983, however, and NASA began to design the first comprehensive search effort mounted thus far.

NASA plans to begin its SETI program in the 1990s with equipment that can scan eight million channels simultaneously over frequencies ranging from one to ten gigahertz *(pages 116-118)*. Using the large probe-tracking radio antennas of the global Deep Space Network and a twenty-acre radio telescope near Arecibo, Puerto Rico, the project combines a surveying sweep of the entire radio sky and targeted studies of nearly 800 Sun-like stars, all within 100 light-years of the Sun. Many astronomers expect NASA SETI to be the most rigorous test to date of the idea that extraterrestrial intelligence exists.

Certainly not everyone is sanguine about that assumption. In a study that began in 1972, American astronomers Ben Zuckerman and Patrick Palmer, both of UCLA, observed more than 600 stars within a radius of seventy-five light-years of the Sun. Over the course of four years, two dozen of them showed some fluctuations in their radio emissions, but nothing that looked like a convincing artificial signal. Zuckerman finally concluded that there was very little chance of finding extraterrestrial intelligence.

The skeptics have a lot of nonsuccess on their side of the discussion. Physicist Enrico Fermi, who with Hungarian-born colleague Leo Szilard achieved the world's first nuclear chain reaction, is credited with best presenting the skeptics' view: Given the number of stars in this and other galaxies, the habitable planets likely to have evolved around them, and the amount of free energy and organic building material available, Fermi mused at Los Alamos in 1950 that many civilizations would inevitably arise and spread out from their home planets. Inevitably, too, they would have been drawn by the beauty and bounty of Earth. "If all this has been happening, they should have arrived here by now," Fermi concluded. "So where are they?" (The droll Szilard is said to have replied, "They are among us, but they call themselves Hungarians.") Frank J. Tipler, one of the main antagonists of the search for extraterrestrial life, views the so-called Fermi paradox as the rock on which the life-search ship founders: The fact that these presumed civilizations have not been detected proves they do not exist *(page 115)*.

DARWIN III

Whether or not intelligence is to be found elsewhere in the galaxy, terrestrial researchers have little recourse but to base their speculations about the nature of that intelligence on an earthly example, whether human, animal, or artificial. One bold thinker in the field is Gerald M. Edelman, director of Rockefeller University's Neurosciences Institute. In 1972 Edelman shared a Nobel prize with Britain's Rodney Porter for discovering the chemical structure of blood's antibody molecule. Afterward, like many Nobel laureates, he changed disciplines, turning to the neurosciences, which draw from a score of chemical and biological fields. In 1988 he published *Neural Darwinism:*

The Theory of Neuronal Group Selection, a difficult book in which a corollary of Charles Darwin's evolution through natural selection—survival of the fittest—is attributed to the brain.

According to Edelman, the human brain's estimated 50 billion nerve cells, called neurons, linked by 100 trillion electrochemical connections, or synapses, make up too complex a structure to have been prefabricated in each individual's genetic messaging system. DNA, Edelman suggests, supplies a rough blueprint but does not prevent the brain from evolving in response to the demands of an infinitely varying world. In neural Darwinism, the brain adapts by altering its electrochemical anatomy, clumping groups of neurons together, and strengthening the synaptic links that permit them to communicate with one another.

The embodiment of the theory is Darwin III, a mathematical model, or set of rules and equations, designed to simulate a simple brain. With the equivalent of about 6,000 neurons and 100,000 or so synapses, Darwin III's thinking apparatus is much less complicated than the brain of an insect. However, when put through its paces by a large computer, the model appears to learn from experience. In thousands of simulations, it has learned to compare what it feels and what it sees to develop categories that it can use to make simple decisions. This learning process is reflected not just in the way Darwin III behaves but in the organization of its numerical brain as well. Where a mathematical simulation of neurons and synapses had originally been a randomly arranged field of brain cells, loosely linked by rather weak connectors, it becomes a highly organized array of tightly clumped neurons where synaptic interconnections steadily strengthen. According to Edelman, what happens in Darwin III is what happens at a vastly more complex level in the human brain: Experience alters the brain's functional anatomy.

Neuroscientists have reached no consensus on Edelman's radical hypothesis, but no one seriously doubts that the brain can change itself or that outside forces can redirect its development. Each brain is thus an individual, despite the structural parameters of its original genetic plan. And what Edelman calls its fluidity—the ability to flash messages between millions of neurons in patterns that respond to external stimuli—suggests that particular brain cells are not dedicated to prescribed duties but may interact in a variety of ways.

Edelman's brain theory holds major implications for life searchers. If intelligence is a dynamic, evolving entity, as biology is, it is as likely to be encountered on other worlds as any other evolutionary product. But given the profound influence of environmental forces, any such intelligence would almost certainly be very unlike its earthly counterpart.

Experiments with Darwin III, a computer simulation of a simple brain, demonstrate that neuronal architecture changes in response to environmental forces. Here, two images of a portion of Darwin III depict the response to a mathematical version of a touch stimulus. Blue lines represent the weakest links between neurons; orange lines, the strongest. First the simulated brain is a random assembly of nerve cells weakly bound by connectors. After repeated touches, some connections have strengthened, and the cells have formed functional groups.

THE UNCOMPUTER

If nothing else, Edelman's work, and parallel investigations by others, serves to caution against making too many assumptions about the nature of intelligence, worldly or otherwise. Three decades ago, for example, researchers attempting to create thinking in machines were convinced that success was near at hand. The ability of digital computers to handle large quantities of material by making many simple computations very quickly seemed to herald a kind of silicon age of thought, an era in which computation would evolve into artificial intelligence. Computers proved so adept at games like chess, which must be played within a fixed framework of rules, that some researchers predicted a computer would soon become the world chess champion.

But no electronic grand masters have appeared. Chess, it turns out, is far more than a merely computational game. According to Hubert Dreyfus, a feisty philosopher at the University of California at Berkeley, and his mathematician brother, Stuart, also at Berkeley, the failure of computers to match the best human chess players suggests that the programmers of the machines have an excessively rigid view of intelligence. Grand masters—whether in chess or other real-life pursuits—do not merely follow the rules.

This insight was borne out in 1979, when the Dreyfus brothers were hired by the Air Force to explore why some jet fighter pilots are better than others. They found that what fliers call "the right stuff" was often not the logical stuff when decisions were made in the life-and-death context of supersonic combat. Pilot interviews revealed that only novices went strictly by the book. As pilots became more proficient, they relied more on context and experience, thinking less about operating a machine than about flying. Victory seemed to go not to computational processes but to intuitive responses.

Hubert Dreyfus delights in being a gadfly to what he calls the "artificial intelligentsia," referring to researchers trying to imbue computers with cognition. In his view, "feelings, memories, and images" must also figure into the intellectual equation. But these aspects are just what scientists are unable to impart to computers, and computers can give back only what has been given them. Science will not be able to duplicate the human brain, Dreyfus has argued, because scientists cannot teach a machine how to feel or give it intuition—which in turn is an amorphous combination of experience, sensory perceptions, and such highly personal qualities as fear, excitement, and the challenge of survival.

Dreyfus is not alone in arguing for the critical role of experience and environment in shaping human intelligence. "Even if we knew all of the genes and what they coded for," Pasko Rakic, a neuroanatomist at the Yale University School of Medicine, told a 1986 meeting of neuroscientists, "we still could not predict in detail what kind of brain you would have." Increasingly, scientists are learning that chemicals, radiation, and other external influences have a major impact on the architecture of the brain, particularly during early development. In a sense, laboratory experiments have begun to reinforce the results obtained by Edelman's rudimentary Darwin III.

BRAIN POWER

The power of environment to shape the brain's anatomy, and thus to drive the evolution of terrestrial intelligence, seems to be written in Earth's fossil record. One day in August 1968, Dale A. Russell, a dinosaur specialist with Canada's National Museum of Natural Sciences in Ottawa, was shown a pile of shattered bones by an Alberta rancher. The bones were those of a wolf-size dinosaur that died on the Canadian prairie 75 million years ago. The creature apparently walked on its hind legs, used a primitive thumb, and had binocular vision. More important, in Russell's view, was the animal's skull. "The cerebral hemispheres were enlarged, exceeding those of any living reptile in relative size and equaling those of some living mammals," he wrote in 1979. "This small dinosaur was a manifestation of the widespread tendency of animate organisms to become more intelligent through geologic time."

Assuming that the ratio of brain to body weight—what is called encephalization—could be used to assess intelligence over evolutionary periods of time, Russell argued that increases in encephalization were governed by environmental forces and that more difficult environments accelerate the brain-growth process. "It is axiomatic," he wrote, "that complicated responses tend to be required of organisms inhabiting complex environments."

Evidence of this came from studies by Russell and others of encephalization rates on the continents of South America and Australia, where the isolation of species seemed to be matched by lower brain-development rates. In other areas of the world, active, foraging creatures tended to develop mental abilities more quickly than organisms that defended themselves passively by such adaptations as becoming poisonous or growing armor. Other environmental factors apparently dictated how such increased brain power was used. Creatures that evolved a high degree of encephalization obviously did not necessarily build technological societies. Whales and dolphins, for example, which have had large brains for more than 25 million years, have not developed technologies. By contrast, the human brain doubled in size only during the past two million years but quickly evolved toward technology. Perhaps more to the point, it also evolved the capacity for language, which many would consider the single most salient aspect of intelligence.

According to Kent State University anthropologist C. Owen Lovejoy, the human brain is predisposed toward communication. Humans are the only creatures able to express the world symbolically and abstractly. Because we are unique in that way, he argues, understanding how language evolved is essential to calculating the likelihood of finding our counterparts elsewhere in the universe.

Again, researchers have turned to the lessons of Earth. Studies have revealed a world that is replete with messages, from the elaborate tap-dancing that bees employ to direct their fellows toward a stand of flowers, to what may be low-frequency dialogues among elephants. Some researchers have managed to communicate across species lines, but only with considerable effort *(pages 111-113).* Their work suggests that it would be no small matter to persuade an alien race to adopt conventions of speech that humans find convenient or simple.

So far, with only a few exceptions, earthbound scientists have done little more than simply listen for a passing radio message from the stars. Efforts to reach out and make contact have been rare and largely symbolic. In 1974 Frank Drake used a gigantic radio telescope at Arecibo, Puerto Rico, to beam an elaborately informative binary-coded message toward a globular star cluster in the constellation Hercules. Now nearing the second decade of its journey at the speed of light, Drake's message will take almost 24,000 years to get there; any answer will take 24,000 more.

Earth has also set adrift the cosmic equivalent of notes in a bottle. *Pioneer 10* and *Pioneer 11,* launched toward the outer planets in 1972 and 1973 and soon to pass into interstellar space, bear plaques that describe the Solar System, Earth, and humans. Similarly, the two Voyager spacecraft launched in the late summer of 1977 and also heading out of the Solar System bear a video recording of Earth sights and sounds, along with a rough map of where and what Earth is *(pages 124-125).*

STARS IN HARNESS

In the view of some scientists, these messages are guilty of what the Soviet Union's Nikolai Kardashev calls terrestrial chauvinism. Kardashev, who conducted pioneering life-search projects in the 1960s and 1970s, contends that SETI programs are wrongly designed around the assumption that alien intelligence is similar to the human version and at about the same stage of technological development. According to Kardashev, the emanations of a society at the present human level would be extremely difficult to detect at interstellar distances beyond about 100 light-years; at that range, the radio emissions from Earth, for example, become so widely spread as to blend into the cosmic background. "Extraterrestrial civilizations have not yet been found," he told an International Astronomical Union meeting in 1985, "because in effect they have not yet been searched for." The search should be modified instead to look for supercivilizations, he said—cosmic societies whose evolution has far surpassed our own and whose works might be more easily seen.

Assuming no restrictions on the scale of intelligent activity—for example, aliens might control their stars' and planets' orbits—Kardashev proposed

four possible scenarios for what he called the urbanization of space and assigned a subjective probability to each. The first, which he gave a 60-percent chance of happening, would be marked by the unification of many planetary homes around many stars, in diverse galaxies spread across one billion to ten billion light-years and ultimately forming a single consolidated object. Not only would their radio signals be detectable, but presumably they would have constructed objects in space that might be discerned against the universal background. Kardashev would search the most powerful known galaxies and quasars, looking for omnidirectional radio emissions at twenty-one centimeters and for beamed signals at a wavelength of 1.5 millimeters, close to the far-infrared wavelengths where heat emitted by engineering structures in space might be detected. Contact would be followed on Earth by preparations to join the higher civilization, incorporating the planet into the gigantic whole.

More as a symbolic gesture than a realistic initiative, several messages proclaiming humanity's existence have been dispatched to the vastness of space. *Pioneers 10* and *11* each bear a plaque *(above, left)* engraved with images of a man and woman, the spacecraft itself, and Earth's position relative to the center of the Milky Way galaxy and fourteen pulsars. On *Voyagers 1* and *2,* a video disk *(above, right)* provides two hours of earthly sights and sounds, including the voices of world leaders and songs of the whale. The gold-plated cover *(above)* diagrams Earth's location and explains how to play the disk.

The second scenario, to which Kardashev assigned a probability of 20 percent, would be unified at the much smaller scale of large clusters of galaxies, spanning distances reckoned in millions of light-years. He would begin searching in the heart of the galaxy cluster in the constellation Virgo and scan the radio sky for signals similar to those postulated for his first scenario. The third possibility, Kardashev theorized, would be unified at the still smaller scale of large galaxies, spanning hundreds of thousands of light-years. Because there has thus far been no evidence of these, he gave them only a 10-percent probability of existing. Still, he would look for them in the nucleus of the Milky Way and at the centers of such large, relatively nearby galaxies as M31 and M33.

His fourth hypothetical society would have completely colonized the universe, but since no signs of them have been detected, they were assigned a zero probability. In the end, he added a fifth category: planetary civilizations not

At right is a graphic representation of a radio message beamed toward a star cluster in the constellation Hercules in 1974. Radio astronomer Frank Drake, guessing that an alien intelligence might understand the binary notation, encoded the numbers one through ten *(top),* followed by the atomic numbers of five elements essential to terrestrial life, the chemical formula of the DNA molecule, and numbers for the average human height and the world's human population. If the receivers decode the message pictorially, they will see images of the human form, the Solar System, and the transmitting radio dish.

unlike Earth's, detectable only by the postapocalyptic flotsam remaining after their technology destroyed them.

According to Kardashev and others, any supercivilization would have to harness first the power of its sun and solar system, then the energy of its galaxy, ultimately spreading until it somehow put the enormous energy of the entire universe to work. These are the kinds of open-ended possibilities favored by Freeman Dyson, a British-born American physicist at Princeton's Institute for Advanced Studies and a gentle promoter of stunning visions of human destiny. Dyson is a slender man who drifts into and out of conversations in midsentence as his agile mind pursues ideas that have, at least superficially, a kind of mad light about them. But he is anything but frivolous. Long a leading theoretical physicist, Dyson was among the midcentury pioneers of quantum electrodynamic theory, the quantum mechanics of particles in magnetic fields.

His broader vision is of a life force undaunted by the deep cold of empty space or by the absence of atmospheric pressure or gravity. If natural selection has not already created a myriad of beings adapted to other worlds, then it will adapt humans to make the ultimate cosmic journey. "I assume," he wrote in 1988, "that life is capable of making itself at home in every corner of the universe, just as it has made itself at home in every corner of this planet." In his view, life is as infinite, or as finite, as the universe itself, which "is like a fertile soil spread out all around us, ready for the seeds of mind to sprout and grow."

On the technological side of this adaptability, Dyson proposed more than twenty years ago that it was only a matter of time before a civilization—on Earth or elsewhere—would begin to remodel its corner of the universe to meet its critical energy needs and compensate for the gradual cooling of its star. In the Solar System, for example, humans, or whatever they will have evolved into by that time, might vaporize the eight other planets and scores of moons into the stuff with which to build a cozy sphere around the Sun. Such a sphere would hold in the energy from the Sun's thermonuclear furnace. From space, these so-called Dyson spheres would appear as dark objects at about room temperature—about the size of what present-day astronomers call brown dwarfs, but a bit colder. On their internal surface, beings of unfathomable dexterity would have built a huge biosphere.

Where Dyson would seek extraterrestrial life by looking for such structures, others would watch for an indirect sign that an alien society had learned to exploit its star—the steady diminution of energy emitted by a distant sun. In 1985 Gerard Bodifee and Camyel de Loore of the Astrophysical Institute Free University of Brussels proposed that the more advanced a society was, the more energy it would use. As more advanced life forms consumed greater amounts of energy, the system would appear from the outside to be suffering a severe case of entropy, the inexorable loss of available energy in the thermodynamic process. As the supersociety transformed its free energy into work and waste, it would leave a kind of trail in space. "An advanced form of life should manifest itself by a large entropy emission," wrote the Belgian team. In other words, high-powered emissions such as x-rays and ultraviolet light would give way to the infrared emissions of waste heat.

Whether or not the sky is home to Dyson spheres or drained stars, it may well be the case that otherworldly intelligences are beyond the comprehension of earthlings. To a species that has hardly scratched the surface of interspecies communication on its own world, an alien civilization could be mature beyond all human imaginings. Extraterrestrial beings might have no space or time references that earthlings can read, no recognizable shape, no voice. They might be immortal by human reckoning, or live and die in mere billionths of a second. Indeed, mortality—of individuals, of societies, stars, and galaxies—may be what precludes humans from actually making contact with an extraterrestrial civilization. Even if the presumably irreconcilable differences between humanity and such creatures could somehow be bridged, the realities of time and space cannot.

Fictional warp drives notwithstanding, the universe appears to be ruled by a 186,000-mile-per-second speed limit—the speed of light traveling through space. According to present theory, this limit cannot even be closely approached by anything possessing mass. And to objects traveling at less than light-speed, space is a moat of formidable proportions. To send a one-way radio message, moving at the speed of light, to the inhabitants of a planet orbiting Epsilon Eridani, for example, would take more than a decade—a long time, granted, but feasible on the human scale. To visit them, however, using a spaceship as fast as Voyager, which can reach 60,000 miles an hour as it flashes past a planet, would require about 100,000 years—many times the duration of human civilization. The same craft would take 350 million years to reach the center of the Milky Way galaxy.

Most scientists expect improvements through technology. Eventually, they believe, humans will build ships capable of reaching an appreciable fraction of the speed of light. At 20 percent of light-speed, Epsilon Eridani is only half a century away, and stars at the hundred-light-year outer limit of NASA's SETI program, a journey of barely 500 years. But as velocities increase, the technological obstacle shrinks before an economic one. Robert Rood of the University of Virginia estimates that, in current monetary terms, the energy needed to accelerate one human to 20 percent of light-speed would cost about two billion dollars, a price that rises sharply as designers add life-support equipment and take into account the inefficiencies of all propulsion systems. With more than one human, the price rises exponentially. Pioneer life searcher Frank Drake has calculated that propelling 100 colonists to a world ten light-years away, at 10 percent of light-speed, would consume the equivalent of 1,000 years of present American energy production. For the moment, interstellar space must remain the province of those imagined masters whose venerable civilizations have learned to harness stars.

GENESIS OF A LIVING PLANET

Life: If it exists, as humans know it, elsewhere in the universe, it most likely is harbored on terrestrial, or Earth-like, planets circling stars similar to the Sun. While no planets have been confirmed outside the Solar System, astronomers have identified many young stars that are more than likely to develop planetary systems. Each is surrounded by a flat disk of orbiting gas and dust, the remnants of material that condensed from a huge cloud and then ignited to create the star itself. Astronomers believe that it is within these so-called protoplanetary disks that planets, including the terrestrial variety, are born.

Terrestrial planets probably grow by the aggregation of solid matter in the dusty midplane of the disk. (Giant gaseous planets like Jupiter and Saturn evolve differently.) Small clouds of dust grains, each of them a microscopic mineral particle, are drawn together by a combination of electrical and chemical attraction, forming denser regions in which the grains collide and stick together. In repeated collisions, the grains grow into tiny pebbles, and the pebbles into rocks, all orbiting the young star. As a rock speeds through the disk, it picks up not only the particles in its path but also, as it grows ever larger, those that are drawn in by its increasing gravitational force. Within a few thousand years after accumulation begins, the dust disk has generated billions of orbiting bodies known as planetesimals, which range in size from boulders to large asteroids, many of them destined for future collisions.

The Era of Impacts

Some ten million years after the birth of a star, the planetesimals in its protoplanetary disk begin colliding like amusement-park bumper cars, merging and growing, attracting neighbors by their stronger gravity. After 100,000 years or so, the biggest planetesimals have become what some experts call "planetary embryos," primitive planets up to a few thousand miles in diameter. Circling the star, each of these huge bodies sweeps up most of the remaining nearby planetesimals, which rain down on its surface and add to its bulk.

Because the nearby objects have been orbiting at about the same velocity and in the same direction as the evolving planet, they hit the surface at low speeds. But occasionally one primitive planet will come close enough to a slightly smaller one to perturb it gravitationally, pulling it into a long elliptical orbit. As the smaller body repeatedly swoops through the disk on this path, it can strike yet another evolving planet at a closing velocity as great as ten miles per second. The result is catastrophic. In the violence of the impact, the colliding bodies are almost entirely melted. In as little as thirty minutes, their dense cores combine, and a terrestrial planet is born.

In a cataclysmic collision *(opposite)*, a Mercury-size planetesimal smashes into one nearly the size of the Earth. The enormous heat of the collision melts both of them, drives off volatile gases, and causes the bodies to merge.

Worlds on a Collision Course

Rocks as large as ten miles across, formed from dust in a protoplanetary disk, revolve around a young star in near-circular orbits. Colliding at low speeds, they merge into larger bodies.

After 100,000 years, many of the rocks have aggregated into planetary embryos, Moon size or larger. Some of them have as much as 70 percent the mass of the Earth.

The gravity of large planetary embryos pulls smaller ones into elliptical orbits that sweep them past other bodies at great speed. Where orbits intersect, violent impacts occur.

AN ATMOSPHERE IS BORN

A young terrestrial planet—its original atmosphere driven off by solar wind and the impact of a planetary embryo that melted the little world—seethes with heat and, in the process, forms a new atmosphere. When the planet was molten, denser material precipitated out and sank toward the center, producing a liquid core of iron and nickel. Lighter materials containing volatile elements such as oxygen and carbon floated above the core and solidified into rock, trapping the gases.

As new impacts from interplanetary debris heat the planet further, gases spew from the interior like steam from a kettle. These gases join with those venting from volcanoes to form a dense, dusty atmosphere. As the millennia pass, heavier particles will drift down from the gaseous shroud. Ultraviolet radiation from the young sun will bathe the remaining gases, causing chemical reactions that produce carbon dioxide and water. Eventually, the residual atmosphere will be hundreds of times less dense, but it may offer a benign environment for the evolution of life.

A molten planet, the product of repeated impacts, is surrounded by a dense atmosphere composed of gases released from the hot interior. Droplets of molten material flung from the planet by the force of the impacts reenter the atmosphere and bombard the surface.

FORMING A PROTECTIVE COVER

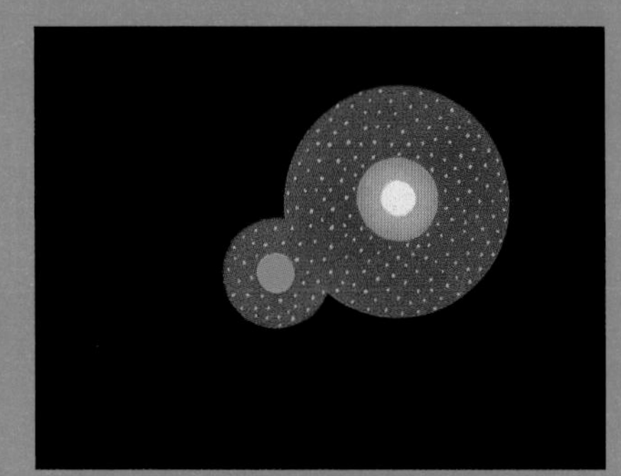

A cutaway diagram shows colliding planetary embryos. Volatile substances *(dots)* are trapped above the dense cores in layers of rock that begin deforming at the moment of impact.

Only fifteen minutes after their initial contact, the cores of the two bodies are changing shape. Much of their volatile matter has been vaporized and driven out by the tremendous heat.

Hours later, gravity has rounded the molten mass back into a sphere. The new core is surrounded by a layer of core matter from the smaller body. Expelled gases form an atmosphere.

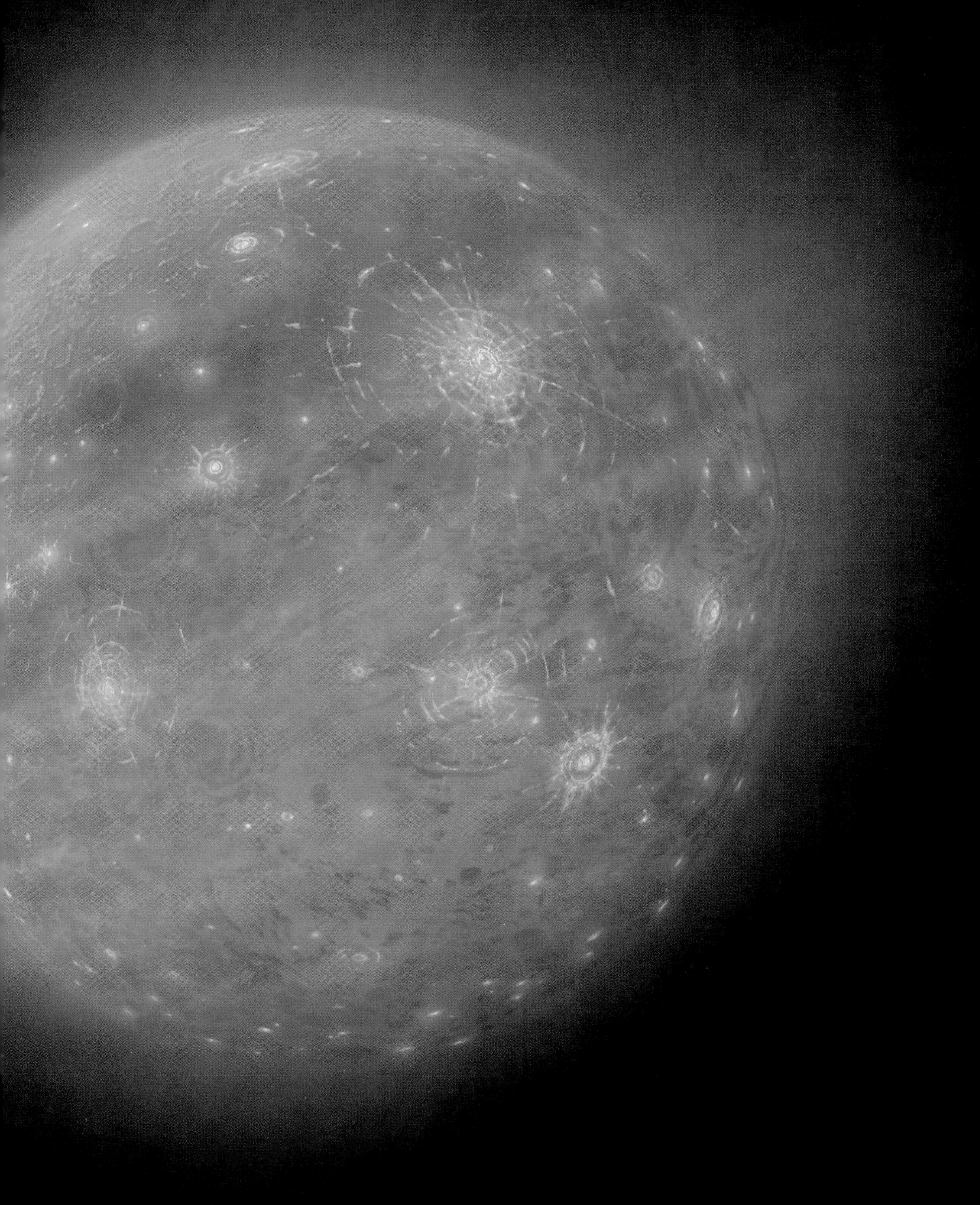

Signs of Life

A terrestrial planet that follows the developmental path taken by Earth will begin to show evidence of life relatively soon after it is fully formed. Alien life would not have to generate radio waves or even possess intelligence in order to signal its presence to earthly astronomers. The life forms would simply have to change their environment in a way that could be detected by large telescopes.

Such a change occurred on the young Earth when primitive, photosynthetic algae liberated oxygen from other gases and pumped it into the atmosphere. Oxygen rising to the upper atmosphere was converted by solar radiation into ozone, a form of oxygen that absorbs ultraviolet radiation and protects life below. Abundant ozone in a distant planet's atmosphere—possible evidence of living organisms—can easily be identified by its characteristic absorption lines in the infrared spectrum of light reflected from the planet.

Astronomers have specified what they need: an infrared telescope that is placed in orbit above the Earth's own ozone shield and that is large enough to pick up the target planet's light in the glare of the adjacent star. A sixteen-meter orbiting instrument would enable them to spot an ozone layer in the atmosphere of a world eighteen light-years away.

A terrestrial planet is transformed by life. Through photosynthesis, primitive plant forms have converted a yellowish atmosphere, laden with noxious sulfur dioxide, into an oxygen-rich blue atmosphere, complete with ozone layer.

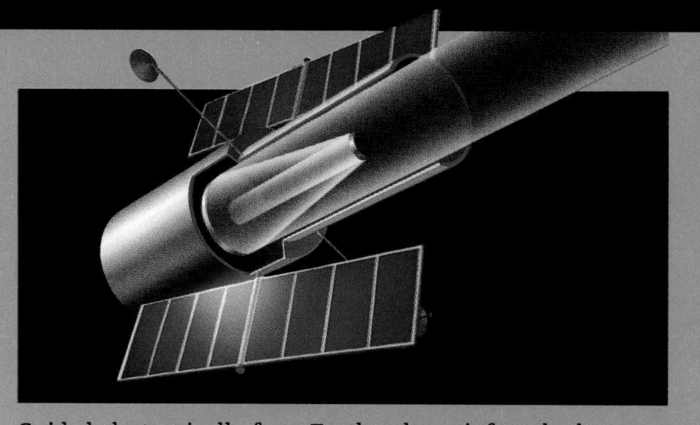

Guided electronically from Earth, a large infrared telescope (cut away to show light focusing on its primary mirror) could search for planets with ozone.

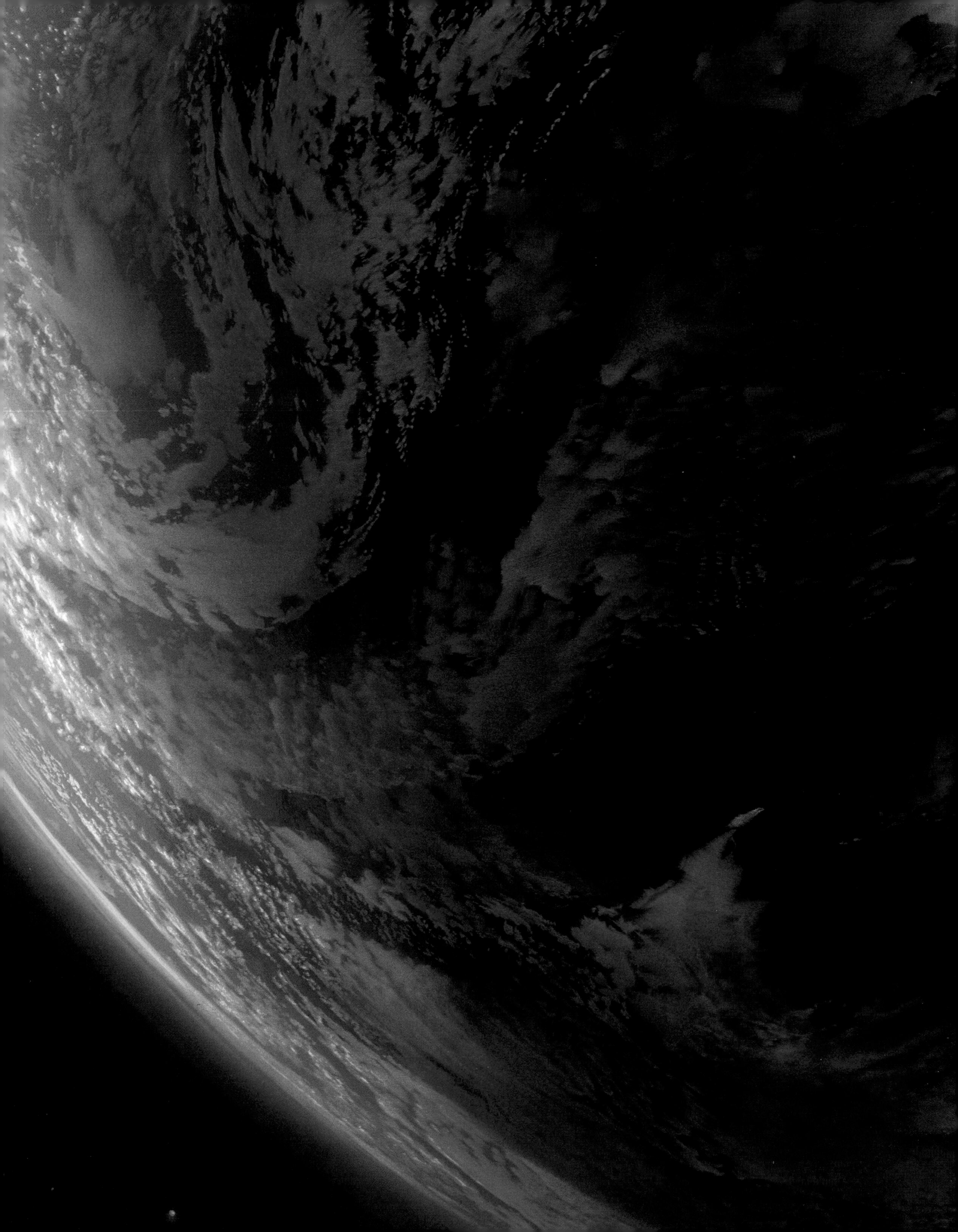

GLOSSARY

Acid: a corrosive substance that releases protons in water. By definition, all acids measure less than 7 on the pH scale. *See* Base.
Acidophilic: adapted to, or requiring, an acidic environment.
Algae: a large and varied group of eucaryotic organisms that carry out photosynthesis but do not have specialized tissue structures.
Alkaline: having the properties of a base. *See* Base.
Amino acid: an organic compound consisting of a standard nine-atom section and a distinctive atomic side chain. Certain kinds of amino acids are the building blocks of proteins.
Ammonia: a chemical compound found in several planetary atmospheres. It is also excreted as a metabolic waste by some terrestrial life forms.
Amphibian: a cold-blooded vertebrate animal adapted to both water and land. Typically, larval amphibians remain in the water and adults live on land.
Anaerobic: adapted to environments without molecular oxygen. Oxygen is often toxic for anaerobic life.
Arc second: a sixtieth of an arc minute, which is in turn a sixtieth of a degree of arc; there are 360 degrees in a circle. Arc seconds, minutes, and degrees measure an object's apparent size and position on the sky.
Asteroid: a small, rocky, airless body that orbits a star.
Astrometry: the science of measuring the motions and positions of stars, planets, and other astronomical bodies.
Atmosphere: a gaseous shell surrounding a planet or other body; also, a unit of pressure equivalent to the atmospheric pressure at sea level on Earth, about 14.7 pounds per square inch.
Bacterium: a member of one of many species of microscopic single-celled organisms.
Base: a chemical compound that reacts with an acid to form a salt. By definition, a base measures above 7 on the pH scale. Base also may refer to a component of DNA or RNA: DNA molecules include the bases adenine, guanine, thymine, and cytosine; RNA substitutes uracil for thymine.
Biochemistry: the branch of chemistry concerned with the chemical processes of living organisms.
Biosphere: the totality of a planet's living things and their habitats.
Brown dwarf: in theory, a dim substellar body with insufficient mass to fuse hydrogen to helium at its core. Brown dwarfs typically have less than one-tenth of a solar mass.
Carbohydrate: an organic compound of carbon, hydrogen, and oxygen atoms. As sugars and starches, carbohydrates are the main source of energy for most terrestrial organisms.
Carbon: a chemical element found in all living matter on Earth and notable for its tendency to form multiple bonds.
Carnivore: a meat-eating animal or plant.
Catalyst: a chemical agent that modifies a reaction but is not itself changed.
Cell: a basic functional and structural unit of living matter. A cell is capable of growth, reproduction, and the excretion of wastes.
Chemosynthesis: the biochemical creation of organic compounds from nonorganic materials using chemical reactions as an energy source.
Chloroplast: an organelle that assists in photosynthesis. Chloroplasts are found in all plant cells.
Circumstellar disk: a large flattened disk of ice, dust, and rock particles suspended in gases orbiting a star. Planets may later form from clumps of this material.
Codon: a unit of genetic information, composed of three nucleotides. Most codons represent a specific amino acid.
Comet: an asteroid-size body of dusty ice that travels in an elongated orbit around the Sun.
Cytoplasm: the liquid material found in the region outside a cell's nucleus.
Deoxyribonucleic acid (DNA): a complex organic compound found in all life on Earth and responsible for the storage of genetic information. A DNA molecule consists of two parallel chains of nucleotides. DNA is named for the sugar deoxyribose, which it contains.
Deuterium: a form of hydrogen having one neutron and one proton in its nucleus. Also known as heavy hydrogen.
Doppler shift: a change in wavelength caused by motion of either the emitter or the observer.
Drake equation: a formula developed by astronomer Frank Drake that estimates the number of Milky Way civilizations capable of communicating with Earth. It is also called the Green Bank equation, after the site of the conference where it was first proposed.
Echo sounder: a device that measures water depth by bouncing sound waves off the bottom and timing their return.
Eclipse: the obscuration of light from a celestial body as it passes through the shadow of another body.
Ecosystem: a community of organisms and their environment.
Electromagnetic radiation: radiation consisting of periodically varying electric and magnetic fields that vibrate perpendicularly to each other and travel through space at 186,000 miles per second.
Electromagnetic spectrum: the array, in order of frequency and wavelength, of electromagnetic radiation, from low-frequency, long-wavelength radio waves to high-frequency, short-wavelength gamma rays.
Electron: a negatively charged particle that normally orbits an atom's nucleus but may exist in isolation.
Element: one of just over 100 substances that cannot be reduced by chemical means to simpler substances.
Endolithic: living within a rocky structure. Most corals and some algae are endolithic.
Endoplasmic reticulum: the membrane structure, extending throughout the outer region of a eucaryotic cell, that transmits substances to and from the nucleus; also, protein synthesis and other reactions take place in this structure.
Energy: the ability to do work, where work is defined as the ability to move mass through space. Life requires energy. An *energy source* adds free energy to a system; an *energy sink* absorbs available free energy.
Entropy: a measure of the energy existing in a system that is unavailable for use.
Enzyme: one of many proteins that accelerate or otherwise affect biochemical reactions without themselves being changed.
Ethane: a colorless, odorless hydrocarbon that is gaseous at Earth temperatures.
Eucaryote: a cell containing a nucleus in which genetic material is stored; also, an organism composed of such cells. The word eucaryote comes from the Greek for "true nucleus."
Evolution: changes over generations in organisms' inheritable characteristics.

Exobiology: the study of, and search for, organisms not native to the Earth. No such organisms are yet known.
Fission: in biology, the asexual reproduction of cells through division into multiple identical cells.
Fix: in biology, to incorporate a naturally occurring element into an organism. Photosynthetic plants fix carbon.
Food chain: the ecological connection among a series of organisms, each of which eats its predecessor in the chain and is eaten by its successor. Photosynthetic or chemosynthetic plants and bacteria begin the chain.
Free oxygen: atmospheric oxygen in its pure form, not bonded in compounds to other elements. On Earth, most free oxygen is produced by photosynthetic plants.
Gaia hypothesis: the theory that the Earth is itself a self-regulating organism, called Gaia after a Greek goddess of the Earth.
Gas chromatograph: a device for the analysis of chemical compounds or mixtures of compounds.
Gene: a complete unit of biochemical information that specifies the series of amino acids needed to make up a particular type of peptide chain, which will in turn form part of a protein.
Golgi apparatus: an organelle, or structure within a cell, that is thought to wrap and dispatch newly constructed proteins.
Gravity: the mutual attraction of separate masses; fundamental force of nature.
Halophilic: adapted to or requiring salt or brine environments.
Homo sapiens: the primate species to which human beings belong, characterized by a brain of about eighty-five cubic inches and by language and tool-making abilities. *Homo sapiens* includes premodern humans such as Neanderthals; present-day humans form a subspecies, *Homo sapiens sapiens.*
Hydrocarbon: one of the large number of organic compounds made up exclusively of carbon and hydrogen atoms. Seven types of hydrocarbons have been detected in space.
Hydrothermal vent: an opening in the ocean floor where water is warmed by escaping heat from the Earth's interior.
Infrared: a band of electromagnetic radiation with a lower frequency and longer wavelength than visible red light. Most infrared radiation is absorbed by the Earth's atmosphere, but certain wavelengths can be detected from Earth.
Infrared detector array: an instrument for the detection and imaging of astronomical infrared signals, consisting of several thousand infrared sensors placed on a small grid.
Infrared excess: infrared radiation detected from certain stars in amounts greater than expected. Infrared excess may indicate the presence of unseen radiating objects such as circumstellar disks or planetesimals.
Infrasonic: pertaining to sounds below the range of human hearing. Certain animal species communicate infrasonically.
Inorganic: pertaining to a chemical compound that does not include both carbon and hydrogen atoms. Also may refer to matter that is not and never has been alive.
Invertebrate: an organism without a backbone.
Ion: an atom that has lost or gained one or more electrons. A neutral atom has an equal number of electrons and protons, giving it a zero net electrical charge. A positive ion of an element has fewer electrons than the neutral atom, a negative ion has more.
Ionosphere: an ionized atmospheric layer. The Earth's ionosphere occurs at altitudes of thirty-five miles and higher.
Isotope: one of two or more forms of a chemical element that have the same number of protons but a different number of neutrons in the nucleus.
Lichen: any of a class of complex organisms consisting of fungi and algae combined in symbiosis. The algae provide nourishment through photosynthesis; the fungi provide structure and water storage.
Light-year: an astronomical distance unit equal to the distance light travels in a vacuum in one year, almost six trillion miles.
Lysosome: an organelle that contains digestive enzymes and defends the cell from intrusive bacteria.
Macrolife: any organism large enough to be seen by the unaided human eye.
Magnetosphere: the region around a planet in which its magnetic field is the dominant magnetic influence.
Mammal: a hairy, warm-blooded vertebrate animal that nurses its young. Almost all mammals bear their young alive, rather than in eggs.
Mass spectrometer: a device used to determine the chemical composition of a substance by measuring the varying masses of the substance's components.
Membrane: a flexible structure that encloses a cell, organelles within a cell, or other tissue. Membranes consist primarily of layered protein and fats.
Metabolism: the biochemical processes that convert energy to a form useful for life.
Meteoroid: a small metallic or rocky body found in space. A meteoroid entering a planet's atmosphere is called a *meteor.* Meteors often burn up in the atmosphere; those that reach the surface are known as *meteorites.*
Methane: a simple five-atom hydrocarbon, gaseous at Earth temperatures and found in the atmospheres of several other Solar System planets.
Microorganism: a living thing too small to be visible to the unaided human eye. Also called a *microbe.*
Microwave window: a range within the radio segment of the electromagnetic spectrum, between 1,000 and 10,000 megahertz. According to theory, high natural background noise in space at other frequencies makes this relatively quiet region the most suitable for interstellar radio communication.
Mitochondrion: an organelle that breaks down chemical compounds for respiration. Animal and plant cells include mitochondria.
Molecule: the smallest unit of an element or compound that retains its properties. A molecule may consist of a single atom or, more commonly, two or more atoms bonded together.
Monomer: *see* Polymer.
Moon: one of a planet's natural satellites, generally no smaller than ten miles in diameter. There are more than fifty known moons in the Solar System.
Multichannel receiver: a device that can simultaneously monitor radio signals on more than one frequency.
Mutation: a random, inheritable change in the genetic pattern of an organism.
Natural selection: the evolutionary process in which well-adapted species survive and increase in numbers while poorly adapted species become extinct.
Neuron: a specialized cell that transmits information

through electrochemical signals. Neurons are distributed throughout a body in a *neural network.*

Neutron: an uncharged particle with a mass similar to a proton's, normally found in the nucleus of an atom.

Nucleic acid: one of the complex organic molecules, including DNA and RNA, that store and transmit genetic information.

Nucleolus: a small body within the cell's nucleus. The function of the nucleolus is not clearly understood but may be related to the synthesis of RNA.

Nucleotide: one of the chemical units that make up a nucleic acid such as DNA or RNA. A nucleotide consists of a phosphate, a sugar, and a base, all bonded together.

Nucleus: a membrane-enclosed structure within a eucaryotic cell that contains genetic material; also, the massive center of an atom, composed of protons and neutrons and orbited by electrons.

Orbit: the path of an object revolving around an astronomical body.

Organelle: a specialized biological structure within a cell, dedicated to a specific function such as respiration or cell defense.

Organic: pertaining to a compound made up of carbon and hydrogen and possibly other elements as well. All life contains organic compounds. The term may also refer to living or once-living material.

Oxidation: a chemical reaction in which oxygen bonds to other elements, typically resulting in the gain or loss of electrons.

Ozone: a three-atom form of oxygen. Earth's stratosphere includes an ozone layer that absorbs dangerous ultraviolet radiation.

Panspermia theory: the hypothesis that life on Earth originated in spores from other life-supporting planets. *Directed panspermia* suggests that the spores were deliberately sent by other intelligences.

Peptide chain: a linear organic compound consisting of up to several hundred amino acids linked together. Proteins are composed of one or more peptide chains.

Permafrost: ground that is permanently frozen unless artificially heated. On Earth, permafrost may extend to a depth of more than 1,000 feet.

Phosphate: a type of inorganic compound. Some phosphates join to sugars and bases to form nucleotides.

Photosynthesis: the biochemical process that converts light to chemical energy and glucose. Photosynthesizing organisms typically consume carbon dioxide and hydrogen and release oxygen as a by-product.

pH scale: a measure of acidity or alkalinity, with values ranging from 0, for extremely acid, to 14, for extremely alkaline. A neutral compound has the pH value 7.

Planet: a large, nonstellar body that orbits a star and shines only with reflected light.

Planetesimal: in theory, an orbiting body as much as six miles across that accretes mass from random collisions and may eventually become a full-scale planet. Planetesimals larger than a few thousand miles in diameter are called *planetary embryos* by some scientists.

Polymer: a compound built out of small, nearly identical chemical units called monomers. Organic polymers include proteins, nucleic acids, carbohydrates, fats, and oils.

Prebiotic: descriptive of complex organic chemicals that may include amino acids but do not include DNA and proteins, the key components of Earth life.

Procaryote: a cell without a nucleus. The term procaryote comes from the Greek for "before the nucleus." *See* Eucaryote.

Protein: one of a class of complex organic molecules necessary to terrestrial life.

Proton: a positively charged particle with about 2,000 times the mass of an electron; normally found in the nucleus of an atom.

Protoplanetary disk: *See* Circumstellar disk.

Pulsar: an astronomical object that emits extremely regular bursts of radio or other energy at intervals of several seconds or less. Pulsars are thought to be spinning neutron stars.

Pyrolysis: the process of heating chemical compounds to separate them for analysis.

Radar: a method of identifying the location or speed of a distant object by bouncing radio waves off its surface and measuring the interval before they return; also, an instrument used for this purpose. The term is an acronym for "radio detection and ranging."

Radio: the least energetic form of electromagnetic radiation, having the lowest frequency and the longest wavelength.

Radio astronomy: the observation and study of radio waves produced by astronomical phenomena.

Radio telescope: an instrument for studying astronomical objects at radio wavelengths.

Reptile: a cold-blooded, vertebrate, nonamphibious animal such as a turtle, lizard, snake, or crocodile.

Ribonucleic acid (RNA): a complex organic molecule named for the sugar ribose, which it contains. Messenger RNA copies genetic information stored in DNA; transfer RNA helps match amino acids to those genetic instructions.

Ribosome: an organelle that synthesizes proteins.

Search for extraterrestrial intelligence (SETI): the scientific field concerned with the search for alien intelligent life; also, the name of a NASA search program.

Silicon: the second most common element, after oxygen, in the Earth's crust. It also makes up seven percent of the matter in the universe.

Sol: the period between one solar transit of a specific point on the sky and the next such transit. The length of a sol varies from planet to planet; on Earth, it is one day.

Solar system: the Sun and its associated system of planets, asteroids, and other orbiting bodies; more generally, any star and the bodies that orbit it.

Sonar: a method that locates and examines underwater objects by transmitting sound waves through the water and analyzing the returning echo; also, an instrument used for this purpose. Sonar is an acronym for "sound navigation ranging."

Species: the basic category of biological classification, consisting of similar organisms capable of interbreeding.

Spectrograph: an instrument that splits light or other electromagnetic radiation into its individual wavelengths and records the result.

Spectroscopy: the study of spectra, including the position and intensity of emission and absorption lines, to learn about the physical processes that create them.

Spectrum: the array of colors or wavelengths obtained by dispersing light from a star or other source, as through a prism. Spectra are often striped with emission or absorption

lines, which can be interpreted to show the chemistry and motion of the light source.
Spontaneous generation: an ancient theory that living organisms may arise from nonliving matter.
Stromatolite: a petrified, stump-shaped fossil consisting of layered algae mats and formed up to three and a half billion years ago.
Sugar: a simple carbohydrate. The sugars ribose and deoxyribose are found in RNA and DNA, respectively.
Symbiosis: the close, interdependent relationship of two unlike organisms.
Synapse: a nerve cell connection point, through which electrochemical impulses are transmitted from one nerve cell to another.
Tardigrade: a tough microorganism capable of surviving in a dormant state under great heat, cold, or aridity, then resuming activity as conditions become more favorable.
Tectonics: the study of a planet's crust, including its structure and processes.
Thermophilic: adapted to very high temperatures, such as those of hot springs or geysers.
Ultraviolet: a band of electromagnetic radiation that has a higher frequency and a shorter wavelength than visible blue light has. Most ultraviolet is absorbed by the Earth's ozone layer, so ultraviolet astronomy is usually performed in space.
Velocity: the speed and direction of motion.
Vertebrate: an organism having a backbone.
Wavelength: the distance from crest to crest or trough to trough of an electromagnetic or other wave. Wavelengths are related to frequency; the longer the wavelength, the lower the frequency.

BIBLIOGRAPHY

Books

Abell, George O., David Morrison, and Sidney C. Wolff, *Exploration of the Universe.* Philadelphia: Saunders College Publishing, 1987.
Alberts, Bruce, et al., *Molecular Biology of the Cell.* New York: Garland, 1983.
Angelo, Joseph A., Jr., *The Extraterrestrial Encyclopedia: Our Search for Life in Outer Space.* New York: Facts On File, 1985.
Asimov, Isaac, *Asimov's Biographical Encyclopedia of Science and Technology.* Garden City, N.Y.: Doubleday, 1982.
Atlas, Ronald M., *Microbiology: Fundamentals and Applications.* New York: Macmillan, 1984.
Attenborough, David, *The Living Planet.* Boston: Little, Brown, 1984.
Baugher, Joseph F.:
On Civilized Stars. Englewood Cliffs, N.J.: Prentice-Hall, 1985.
The Space-Age Solar System. New York: John Wiley & Sons, 1988.
Beatty, J. Kelly, Brian O'Leary, and Andrew Chaikin, eds., *The New Solar System.* Cambridge, Mass.: Sky, 1982.
Bernal, J. D., *The Origin of Life.* Cleveland, Ohio: World, 1967.
Berrill, Jacquelyn, *Wonders of How Animals Learn.* New York: Dodd, Mead, 1979.
Billingham, John, ed., *Life in the Universe.* Cambridge, Mass.: MIT Press, 1982.
Briggs, Geoffrey, and Fredric Taylor, *The Cambridge Photographic Atlas of the Planets.* Cambridge: Cambridge University Press, 1986.
Brock, Thomas D., and M. Louise Brock, *Life in the Geyser Basins.* Washington, D.C.: U.S. Department of Interior, 1971.
Cairns-Smith, A. G., *Seven Clues to the Origin of Life.* Cambridge: Cambridge University Press, 1985.
Cairns-Smith, A. G., and H. Hartman, eds., *Clay Minerals and the Origin of Life.* Cambridge: Cambridge University Press, 1986.
Calder, Nigel, *Timescale.* New York: Viking Press, 1983.
Crail, Ted, *Apetalk and Whalespeak.* Los Angeles: J. P. Tarcher, 1981.
Crowe, Michael J., *The Extraterrestrial Life Debate, 1750-1900.* Cambridge: Cambridge University Press, 1986.
Curtis, Helena, *Biology.* New York: Worth, 1979.
Day, William, *Genesis on Planet Earth: The Search for Life's Beginning.* New Haven, Conn.: Yale University Press, 1984.
Dyson, Freeman:
Infinite in All Directions. New York: Harper & Row, 1988.
Origins of Life. Cambridge: Cambridge University Press, 1987.
Ezell, Edward Clinton, and Linda Neuman Ezell, *On Mars: Exploration of the Red Planet, 1958-1978.* Washington, D.C.: National Aeronautics and Space Administration, 1984.
Feinberg, Gerald, and Robert Shapiro, *Life beyond Earth.* New York: William Morrow, 1980.
Frank, Claus Jürgen, ed., *Wonders of Nature.* New York: Macmillan, 1980.
Freundlich, Martin M., and Bernard M. Wagner, eds., *Exobiology: The Search for Extraterrestrial Life.* Vol. 19 of *AAS Science and Technology Series.* Washington, D.C.: American Astronautical Society, 1969.
Gamlin, Linda, and Gail Vines, eds., *The Evolution of Life.* New York: Oxford University Press, 1987.
Gibor, Aharon, *Conditions for Life.* San Francisco: W. H. Freeman, 1976.
Gillispie, Charles Coulston, ed., *Dictionary of Scientific Biography.* Vols. 1-15. New York: Charles Scribner's Sons, 1980.
Goldsmith, Donald, and Tobias Owen, *The Search for Life in the Universe.* Menlo Park, Calif.: Benjamin/Cummings, 1980.
Griffin, Donald R.:
Animal Thinking. Cambridge, Mass.: Harvard University Press, 1984.
The Question of Animal Awareness. New York: Rockefeller University Press, 1981.
Grzimek, Bernhard, ed.:
Grzimek's Animal Life Encyclopedia. New York: Van

Nostrand Reinhold, 1974.
Grzimek's Encyclopedia of Evolution. New York: Van Nostrand Reinhold, 1976.

Hartmann, William K., *Moons and Planets.* Belmont, Calif.: Wadsworth, 1983.

Hochachka, Peter W., and Michael Guppy, *Metabolic Arrest and the Control of Biological Time.* Cambridge, Mass.: Harvard University Press, 1987.

Hochachka, Peter W., and George N. Somero, *Biochemical Adaptation.* Princeton, N.J.: Princeton University Press, 1984.

Horowitz, Norman H., *To Utopia and Back: The Search for Life in the Solar System.* San Francisco: W. H. Freeman, 1986.

Hoyle, Fred, and Chandra Wickramasinghe, *Lifecloud.* New York: Harper & Row, 1978.

Hoyt, William Graves, *Lowell and Mars.* Tucson: University of Arizona Press, 1976.

Jackson, Joseph H., and John H. Baumert, *Pictorial Guide to the Planets.* New York: Harper & Row, 1981.

Kellermann, K. I., and G. A. Seielstad, eds., *The Search for Extraterrestrial Intelligence.* Dordrecht, Netherlands: D. Reidel, 1985.

Kuiper, Gerard P., *The Atmospheres of the Earth and Planets.* Chicago: University of Chicago Press, 1952.

Kutter, G. Siegfried, *The Universe and Life: Origins and Evolution.* Boston: Jones and Bartlett, 1987.

Laustsen, Svend, Claus Madsen, and Richard M. West, *Exploring the Southern Sky.* Berlin: Springer-Verlag, 1987.

Lovelock, James, *The Ages of Gaia.* New York: W. W. Norton, 1988.

Lowell, Percival, *Mars and Its Canals.* New York: Macmillan, 1906.

McDonough, Thomas R., *The Search for Extraterrestrial Intelligence.* New York: John Wiley & Sons, 1987.

Mamikunian, G., and M. H. Briggs, eds., *Current Aspects of Exobiology.* Oxford: Pergamon Press, 1965.

Margulis, Lynn, *Early Life.* Boston: Science Books International, 1982.

Margulis, Lynn, and Dorion Sagan, *Microcosmos.* New York: Summit Books, 1986.

Margulis, Lynn, and Karlene V. Schwartz, *Five Kingdoms.* New York: W. H. Freeman, 1988.

Marten, Michael, John May, and Rosemary Taylor, *Weird & Wonderful Wildlife.* San Francisco: Chronicle Books, 1983.

Miller, Ron, and William K. Hartmann, *The Grand Tour: A Traveler's Guide to the Solar System.* New York: Workman, 1981.

Murray, Bruce, Michael C. Malin, and Ronald Greeley, *Earthlike Planets.* San Francisco: W. H. Freeman, 1981.

Oparin, A. I., *The Origin of Life.* New York: Macmillan, 1938.

Papagiannis, Michael D., ed., *The Search for Extraterrestrial Life: Recent Developments.* Dordrecht, Netherlands: D. Reidel, 1985.

Patterson, Francine, *Koko's Kitten.* New York: Scholastic, 1985.

Peters, Roger, *Mammalian Communication: A Behavioral Analysis of Meaning.* Monterey, Calif.: Brooks/Cole, 1980.

Ponnamperuma, Cyril, *The Origins of Life.* New York: E. P. Dutton, 1972.

Ponnamperuma, Cyril, ed., *Comets and the Origin of Life.* Dordrecht, Netherlands: D. Reidel, 1981.

Ponnamperuma, Cyril, and Lynn Margulis, eds., *Limits of Life.* Dordrecht, Netherlands: D. Reidel, 1980.

Preiss, Byron, ed., *The Universe.* New York: Bantam Books, 1987.

Reader, John, *The Rise of Life: The First 3-5 Billion Years.* London: Collins, 1986.

Regis, Edward, Jr., ed., *Extraterrestrials.* Cambridge: Cambridge University Press, 1985.

Rinard, Judith E., *Dolphins: Our Friends in the Sea.* Washington, D.C.: National Geographic Society, 1986.

Robison, Bruce H., *Lurkers of the Deep.* New York: David McKay, 1978.

Rood, Robert T., and James S. Trefil, *Are We Alone?* New York: Charles Scribner's Sons, 1981.

Sagan, Dorion, and Lynn Margulis, *Garden of Microbial Delights: A Practical Guide to the Subvisible World.* Boston: Harcourt Brace Jovanovich, 1988.

Sattler, Helen Roney, *The Illustrated Dinosaur Dictionary.* New York: Lothrop, Lee & Shepard Books, 1983.

Shklovskii, I. S., and Carl Sagan, *Intelligent Life in the Universe.* San Francisco: Holden-Day, 1966.

Simpson, George Gaylord, *Fossils and the History of Life.* New York: Scientific American Books, 1983.

Smoluchowski, Roman, *The Solar System.* New York: Scientific American Books, 1983.

Stanley, Steven M., *Extinction.* New York: Scientific American Books, 1987.

Stebbins, G. Ledyard, *Darwin to DNA, Molecules to Humanity.* New York: W. H. Freeman, 1982.

Trefil, James S., *Space Time Infinity.* New York: Pantheon Books, 1985.

Viking Lander Imaging Team, *The Martian Landscape.* Washington, D.C.: National Aeronautics and Space Administration, 1978.

Villee, Claude A., Warren F. Walker, Jr., and Robert D. Barns, *General Zoology.* Philadelphia, Saunders College Publishing, 1984.

Periodicals

Ballard, Robert D., and J. Frederick Grassle, "Return to Oases of the Deep." *National Geographic,* November 1979.

Brock, Thomas D., "Life at High Temperatures." *Science,* November 1967.

Brownlee, Shannon, "Bizarre Beasts of the Abyss." *Discover,* July 1984.

Cairns-Smith, A. G., "The First Organisms." *Scientific American,* June 1985.

Campbell, Bruce, "Planets around Other Stars." *The Planetary Report,* May/June 1988.

Ciaccio, Edward J., "Atmospheres." *Astronomy,* May 1984.

Corbliss, John B., and Robert D. Ballard, "Oases of Life in the Cold Abyss." *National Geographic,* October 1977.

Crick, Francis, "The Seeds of Life." *Discover,* October 1981.

Croswell, Ken, "Titan: Slumbering Giant." *Space World,* January 1988.

Crowley, Thomas J., and Gerald R. North, "Abrupt Climate Change and Extinction Events in Earth Histo-

ry." *Science,* May 20, 1988.
Cruikshank, Dale P., "20th-Century Astronomer." *Sky & Telescope,* March 1974.
Cruikshank, Dale P., and Jerome Apt, "Methane on Triton: Physical State and Distribution." *Icarus 58,* 1984.
Cruikshank, Dale P., Robert Hamilton Brown, and Roger N. Clark, "Nitrogen on Triton." *Icarus,* 1984, pages 293-305.
DeVincenzi, Donald L., "Life in the Universe." *Planetary Report,* November/December 1987.
Dickerson, Richard E., "Chemical Evolution and the Origin of Life." *Scientific American,* September 1978.
Dillman, Terry, "Death in Frozen Wasteland." *The Rochester Times-Union,* December 12, 1973.
Easterbrook, Gregg, "Are We Alone?" *Atlantic,* August 1988.
Evans, J. E., and E. Walter Maunder, "Experiments as to the Actuality of the 'Canals' Observed on Mars." *Monthly Notices of the Royal Astronomical Society,* June 1903.
Ferris, James P., "The Origin of Life." *Planetary Report,* November/December 1987.
Friedmann, E. Imre, "Endolithic Microorganisms in the Antarctic Cold Desert." *Science,* February 26, 1982.
Friedmann, E. Imre, and Rebecca Weed, "Microbial Trace-Fossil Formation, Biogenous, and Abiotic Weathering in the Antarctic Cold Desert." *Science,* May 8, 1987.
Greenstein, George, "An Invitation to Strangers." *Science 85,* June 1985.
Hellerstein, David, "Plotting a Theory of the Brain." *New York Times Magazine,* May 22, 1988.
Horowitz, N. H., "Mars, Then and Now." *Planetary Report,* November/December 1987.
Horowitz, N. H., Roy E. Cameron, and Jerry S. Hubbard, "Microbiology of the Dry Valleys of Antarctica." *Science,* April 21, 1972.
"Images: A New Perspective on Mars." *Sky & Telescope,* June 1987.
"Images: An Orbital View of Mars." *Sky & Telescope,* August 1988.
Irvine, William M., "Chemistry between the Stars." *Planetary Report,* November/December 1987.
Jannasch, Holger W., "Chemosynthesis: The Nutritional Basis for Life at Deep-Sea Vents." *Oceanus,* fall 1984.
Johnstone, Bill, "New Worlds: A Night with the Canada-France-Hawaii Telescope." *Rotunda,* summer 1988.
Kerr, Richard A., "Searching Land and Sea for the Dinosaur Killer." *Science,* August 21, 1987.
Khare, B. N., et al., "The Organic Aerosols of Titan." *Life Sciences and Space Research XXI(2).* Oxford: Pergamon Press, 1984.
Klein, Harold P., "Biology and the Exploration of Mars." *Planetary Report,* November/December 1987.
Lake, James A., "The Ribosome." *Scientific American,* August 1981.
Langone, John, "Cyril Ponnamperuma: Meteorites and the Stuff of Life." *Discover,* November 1983.
Lightman, Alan, "E.T. Call Harvard." *Science 85,* September 1985.
McKay, Christopher P., "Terraforming: Making an Earth of Mars." *Planetary Report,* November/December 1987.
McKean, Kevin, "Life on a Young Planet." *Discover,* March 1983.
Monastersky, Richard, "The Plankton-Climate Connection." *Science News,* December 5, 1987.
Olson, Edward C., "Intelligent Life in Space." *Astronomy,* July 1985.
Owen, Tobias, "Time Travel and Chemical Evolution: A Look at the Outer Solar System." *Planetary Report,* November/December 1987.
Papagiannis, Michael D., "Bioastronomy: The Search for Extraterrestrial Life." *Sky & Telescope,* June 1984.
Parlour, Richard, "Undesirable Aliens." *Astronomy,* July 1984.
Patterson, Francine, "Conversations with a Gorilla." *National Geographic,* October 1978.
Pellegrino, Charles R., "Oceans Above." *Final Frontier,* August 1988.
Ponnamperuma, Cyril, and Peter Molton, "Life on Jupiter?" *New Scientist,* December 6, 1973.
Rensberger, Boyce, "Life in Limbo." *Science 80,* November 1980.
Reynolds, Ray T., et al., "On the Habitability of Europa." *Icarus 56,* 1983.
Rona, Peter A., "Metal Factories of the Deep Sea." *Natural History,* January 1988.
Rose, Frank, "The Black Knight of AI." *Science,* March 1985.
Sagan, Carl, and E. E. Salpeter, "Particles, Environments, and Possible Ecologies in the Jovian Atmosphere." *Astrophysical Journal Supplement Series,* December 1976.
Scattergood, T., D. Des Marais, and L. Jahnke, "Life's Origin: The Cosmic, Planetary and Biological Processes." *Planetary Report,* November/December 1987.
Schusterman, Ronald J., "Artificial Language Comprehension in Dolphins and Sea Lions: The Essential Skills." *Psychological Record,* 1988, pages 311-348.
Serpell, James, "A Horse Named Hans." *Courier,* February 1988.
Smith, Bradford A., and Richard T. Terrile, "A Circumstellar Disk around Pictoris." *Science,* December 21, 1984.
Somero, George N., "Physiology and Biochemistry of the Hydrothermal Vent Animals." *Oceanus,* fall 1984.
Spangenburg, Ray, and Diane Moser, "The Loneliest Place on Earth." *Final Frontier,* April 1988.
Squyres, Steven W., and Ray T. Reynolds, "The Solar System's Other Ocean." *Planetary Report,* May/June 1983.
Stoeckenius, Walther, "The Purple Membrane of Salt-Loving Bacteria." *Scientific American,* June 1976.
Trachtman, Paul, "The Search for Life's Origins—and a First 'Synthetic Cell.' " *Smithsonian,* June 1984.
Vogel, Shawna, "E.T., Phone NASA." *Discover,* October 1987.
Waldrop, M. Mitchell, "Machinations of Thought." *Science,* March 1985.
Weisburd, Stefi, "Death-Defying Dehydration." *Science News,* February 13, 1988.
Wharton, Robert A., Jr., "Gathering Evidence: The Case for Past Life on Mars." *Space World,* September 1988.
Woese, Carl R., "Archaebacteria." *Scientific American,* June 1981.
Wolkomir, Richard, "The Wizard of Ooze." *Omni,* January 1985.

Zahl, Paul A., "Life in a 'Dead' Sea—Great Salt Lake." *National Geographic,* August 1967.

Other Publications

Morrison, Philip, John Billingham, and John Wolfe, eds., "The Search for Extraterrestrial Intelligence" (NASA SP-419). Washington, D.C.: National Aeronautics and Space Administration, Scientific and Technical Information Office, 1977.
"SETI: Search for Extraterrestrial Intelligence." *JPL Fact Sheet.* Pasadena, Calif.: National Aeronautics and Space Administration, Jet Propulsion Laboratory, June 5, 1985.
"Voyager: Saturn Science Summary." Pasadena, Calif.: National Aeronautics and Space Administration, Jet Propulsion Laboratory, no date.
"Voyager Encounters Jupiter." Washington, D.C.: National Aeronautics and Space Administration, July 1979.
"Voyager 2 at Uranus." *JPL Fact Sheet.* Pasadena, Calif.: National Aeronautics and Space Administration, Jet Propulsion Laboratory, no date.

INDEX

Numerals in italics indicate an illustration of the subject mentioned.

ACKNOWLEDGMENTS

The editors of *Life Search* wish to thank these people for their contributions: Michael F. A'Hearn, University of Maryland, College Park; John Appleby, Jet Propulsion Laboratory, Pasedena, Calif.; Dana Backman, Kitt Peak National Observatory, Tucson, Ariz.; Bruce T. E. Campbell, University of Victoria, Victoria, Canada; Michael A. Cunningham, Ohio State University, Columbus; Robert J. Emry, Smithsonian Institution, Washington, D.C.; E. Imre Friedmann, Florida State University, Tallahassee; Einar Gall, Neurosciences Institute, New York; Louis Herman, University of Hawaii at Manoa, Honolulu; Nicholas Hotton III, Smithsonian Institution, Washington, D.C.; James Kaler, University of Illinois, Urbana; Heidi Klein, Bildarchiv Preussischer Kulturbesitz, West Berlin; Christopher P. McKay, NASA—Ames Research Center, Moffett Field, Calif.; Sten Odenwald, Naval Research Laboratory, Washington, D.C.; Peter A. Rona, Atlantic Oceanographic and Meteorological Laboratories, Miami, Fla.; Robert T. Rood, University of Virginia, Charlottesville; Raymond T. Rye II, Smithsonian Institution, Washington, D.C.; Ronald J. Schusterman, Gorilla Foundation, Woodside, Calif.; Larry Seyfarth, University of Pennsylvania, Philadelphia; George N. Somero, Scripps Institution of Oceanography, La Jolla, Calif.; Steven Squyres, Cornell University, Ithaca, N.Y.; Alan Stern, University of Colorado, Boulder; Reid Thompson, Cornell University, Ithaca, N.Y.; David M. Ward, Montana State University, Bozeman; Richard M. West, European Southern Observatory, Garching, West Germany; George Wetherill, Department of Terrestrial Magnetism, Carnegie Institution, Washington, D.C.

PICTURE CREDITS

The sources for the illustrations in this book are listed below. Credits from left to right are separated by semicolons; credits from top to bottom are separated by dashes.

Cover: Art by Damon M. Hertig. Front and back endpapers: Computer-generated art by John Drummond. 2, 3: Art by Damon M. Hertig, background, European Southern Observatory, Garching, West Germany. 8, 9: Courtesy *Newton* magazine. 10: Detail from pages 8, 9. 12: Courtesy Lowell Observatory Photographs. 13: From *Mars and Its Canals,* by Percival Lowell, MacMillan Company, New York, 1906, copied by Larry Sherer. 15: From *Monthly Notices of the Royal Astronomical Society,* Volume 63, courtesy the Royal Astronomical Society, London. 17: Courtesy Lowell Observatory Photographs. 20, 21: Art by Paul Hudson; art by Nick Schrenk. 22-24: Art by Nick Schrenk. 26, 27: Alfred McEwen, U.S. Geological Survey, Flagstaff, Ariz. 29: Detail from pages 38, 39. 30-39: Art by Yvonne Gensurowsky of Stansbury, Ronsaville and Wood Inc. 40, 41: Courtesy Professor Walter D. Keller, Department of Geology, University of Missouri-Columbia. 42: Detail from pages 40, 41. 43: Dennis Anderson/*Astronomy* magazine. 44: Courtesy Dr. Michael A'Hearn, University of Maryland—art by Yvonne Gensurowsky of Stansbury, Ronsaville and Wood Inc. 45: Courtesy Dr. David Deamer, Department of Zoology, University of California at Davis—art by Yvonne Gensurowsky of Stansbury, Ronsaville and Wood Inc. 47-63: Art by Damon M. Hertig. 66, 67: Doug Allan/Oxford Scientific Films, Long Hanborough, Oxfordshire; Hans Pfletschinger/Peter Arnold; art by Damon M. Hertig—Dr. E. Imre Friedmann, Professor of Biology, Director, Polar Desert Research Center, Florida State University; R. Schuster, Department of Entomology, University of California at Davis. 68: Peter David/Planet Earth Pictures, London. 69: Art by Damon M. Hertig—George N. Somero—Larry Malin/Planet Earth Pictures, London. 70, 71: Art by Damon M. Hertig (2)—Robert Hessler/Planet Earth Pictures, London; Holger Jannasch, Woods Hole Oceanographic Institution; Kim Taylor/Bruce Coleman; Paul A. Zahl © National Geographic Society. 72, 73: Michael Freeman/Bruce Coleman; art by Damon M. Hertig—David M. Ward, Department of Microbiology, Montana State University. 74, 75: Jet Propulsion Laboratory, Pasadena, Calif.; background, Dennis di Cicco/*Sky & Telescope.* 76: Detail from pages 74, 75. 77: Courtesy Dr. Dana Backman and Steve Rooke, National Optical Astronomy Observatories. 78: Jim Richardson/West Light. 80, 81: Institute for Astronomy, University of Hawaii. 82, 83: Courtesy NASA. 84, 85: Courtesy Yerkes Observatory, University of Chicago. 86, 87: Art by Nick Schrenk. 88: Painting by Ghesley Bonestell from *The Conquest of Space,* by Willy Ley, Viking Press, New York, 1949. 90-93: Art by Matt McMullen. 96-105: Art by Damon M. Hertig. 106, 107: Courtesy Barrett Gallagher. 108: Detail from pages 106, 107. 109: Courtesy Professor Melvin Calvin. 111: Background art by Stephen R. Wagner. Cube, clockwise from top: Dr. Ronald H. Cohn/The Gorilla Foundation; © 1988 by Sea World, Inc., reproduced by permission; Bildarchiv Preussischer Kulturbesitz, West Berlin. 112: Background art by Stephen R. Wagner. Cube, clockwise from top: Dr. Diana Reiss, courtesy Marine World; Ann Rabushka/Woods Hole Oceanographic Institution; courtesy Lawrence S. Burr. Background art by Stephen R. Wagner. Cubes, clockwise from top: Hans Pfletschinger/Peter Arnold; Darryl W. Bush, Marine World Africa USA; courtesy Dr. Michael A. Cunningham—Dr. Ronald H. Cohn/The Gorilla Foundation (2); courtesy David Carter; (right) Alan Levenson, Kewalo Basin Marine Mammal Laboratory; courtesy Dr. Ronald J. Schusterman, University of California at Santa Cruz. 115: Courtesy Tulane University. 116-118: Art by Sam Ward. 120, 121: Courtesy Neurosciences Institute. 124, 125: Courtesy NASA; courtesy NASA/JPL—courtesy Arecibo Observatory/NAIC/Cornell University/NSF. 128-133: Art by Paul Hudson.

VOYAGE THROUGH THE UNIVERSE

SERIES DIRECTOR: Roberta Conlan
Series Administrator: Judith W. Shanks

Editorial Staff for *Life Search*
Designer: Ellen Robling
Associate Editor: Blaine Marshall (pictures)
Text Editors: Carl A. Posey (principal), Pat Daniels
Researchers: Patti H. Cass, Karin Kinney, Edward O. Marshall
Writer: Esther Ferington
Assistant Designer: Barbara M. Sheppard
Editorial Assistant: Jayne A. L. Dover
Copy Coordinator: Darcie Conner Johnston
Picture Coordinator: Richard Karno

Special Contributors: Susan Bender, Deborah Blum, Lee Dye, Ken Frasier, Peter Gwynne, Leon Jaroff, Steve Maran, Jim Merritt, Chuck Smith (text); Vilasini Balakrishnan, Andrea Corell, Sanjoy Ghosh, Jocelyn Lindsay, Tom Sodroski, Carolyn Tozier (research); Barbara L. Klein (index)

CONSULTANTS

DAVID W. DEAMER is a professor in the Department of Zoology at the University of California at Davis. His research interest is organic substances found in meteorites and their possible role in the origin of cellular structures on the early Earth.

HAROLD KLINE, for many years the Director of Life Sciences at NASA's Ames Research Center, is chairman of the Committee on Planetary Biology and Chemical Evolution for the U.S. Space Science Board of the National Academy of Sciences. His special research area is microbial biology.

G. SIEGFRIED KUTTER, an astrophysicist with an extensive background in the biological sciences, is on leave from the Evergreen State College to the National Science Foundation's Division of Astronomical Sciences.

THOMAS R. MCDONOUGH is a lecturer in engineering at the California Institute of Technology and coordinator of the search for extraterrestrial intelligence at the Planetary Society.

TOBIAS OWEN is professor of astronomy at the State University of New York at Stony Brook. One of his areas of expertise is planets and planetary atmospheres.

DIANA REISS, an expert in interspecies communication, consults for the SETI project of NASA's Ames Research Center. She is employed by Marine World.

BRADFORD A. SMITH, a professor of planetary sciences at the University of Arizona, heads the Voyager scientific imaging team for NASA.

GERALD A. SOFFEN, associate director of Earth and Space Sciences, NASA-Goddard Space Flight Center, served as project director of the Viking missions.

Library of Congress Cataloging in Publication Data
Life Search/by the editors of Time-Life Books.
p. cm. (Voyage through the universe).
Bibliography: p.
Includes index.
ISBN 0-8094-6866-2
ISBN 0-8094-6867-0 (lib. bdg.)
1. Life on other planets. I. Time-Life Books.
II. Series.
QB54.L485 1989
574.999—dc19 88-39253 CIP

For information on and a full description of any of the Time-Life Books series, please call 1-800-621-7026 or write:
Reader Information
Time-Life Customer Service
P.O. Box C-32068
Richmond, Virginia 23261-2068

Third printing 1990. Printed in U.S.A.
Published simultaneously in Canada.
School and library distribution by Silver Burdett Company, Morristown, New Jersey 07960.

Earth: diameter 7,926 miles

Neptune: diameter 30,700 miles

Uranus: diameter 31,600 miles

Red supergiant: diameter 400 million miles

Solar System: diameter 7.5 billion miles

Globular cluster: diameter 2×10^{14} miles

Milky Way: diameter 100,000 light-years

Local Group of galaxies:
6 million light-years across

Largest double radio source:
length 17 million light-years